"I Am Destroying the Land!"

Conflict and Social Change Series

Scott Whiteford and William Derman, Series Editors
Michigan State University

"I Am Destroying the Land!"

The Political Ecology of Poverty and Environmental Destruction in Honduras

Susan C. Stonich

LONDON AND NEW YORK

Conflict and Social Change Series

Photos courtesy of Susan C. Stonich

First published 1993 by Westview Press

Published 2020 by Routledge
52 Vanderbilt Avenue, New York, NY 10017
2 Park Square, Milton Park, Abingdon, Oxon OX14 4RN

Routledge is an imprint of the Taylor & Francis Group, an informa business

Library of Congress Cataloging-in-Publication Data
Stonich, Susan C.
  "I am destroying the land!" : the political ecology of poverty and environmental destruction in Honduras / Susan C. Stonich.
    p. cm. — (Conflict and social change series)
  Includes bibliographical references.
  ISBN 0-8133-8649-7
    1. Honduras—Economic conditions. 2. Agriculture—Environmental aspects—Honduras. 3. Rural poor—Honduras. 4. Environmental degradation—Honduras. 5. Economic development—Environmental aspects. I. Title. II. Series.
HC145.Z9E57 1993
363.7'0097283—dc20   93-2258
                                                                        CIP

ISBN 13: 978-0-367-01120-8 (hbk)
ISBN 13: 978-0-367-16107-1 (pbk)

To Matt and Aaron

Contents

7   **Conclusions: The Political Ecology
    of Development**   161

Illustrations

Figures

Photos

Acknowledgments

This book integrates information from many disciplines and was facilitated by many colleagues and friends over a period of more than a decade. I owe a very special debt of gratitude to the most important collaborators, the several hundred peasant women and men who cooperated with me and befriended my family. Much of the book is written from their perspective. My hope is that the work adequately conveys their courage and ingenuity. I owe special thanks to all my friends in Honduras and to my family who have supported and encouraged me despite significant inconveniences to themselves. I owe a great deal to Billie DeWalt, who directed my Ph.D. dissertation, which was the starting point for this study. I recognize the importance of the earlier work done in the region by Robert White, Bill Durham, and Jeff Boyer, which provided the intellectual foundation for my own studies, as did the work of Eric Wolf, John Bennett, Emilio Moran, Bonnie McCay, Robert Netting, and the many other anthropologists who have attempted to integrate "the environment" into social science theory and methodology.

The University of Kentucky provided funding for an initial period of fieldwork in 1981 through a Summer Research Fellowship and again in 1984 for training and use of the Geographic/Agronomic Information System developed by the Comprehensive Resource Inventory Evaluation System (CRIES) project at Michigan State University. Fieldwork from 1982 through 1984 was funded by the International Sorghum Millet Collaborative Research Support Project (INTSORMIL), which subsequently funded shorter periods of fieldwork and my attendance at conferences in Honduras in 1986 and 1987. The Council for the International Exchange of Scholars provided a Fulbright Senior Research Award for 1989–1990 that was crucial for the study. The staff of the Fulbright office in Tegucigalpa was helpful in every way that they could be. The University of California at Santa Barbara granted a quarter leave of absence, which allowed me to extend the period of that fieldwork; two Faculty Career Development Awards for release time in 1990 and 1992 for data analysis and manuscript preparation; and an Academic Senate Research Award in 1990–1991 for a preliminary study of the effects of shrimp mariculture and other nontraditional exports in the south.

Throughout the decade, the staff of the Honduran Institute of Anthropology and History (IHAH), especially Gloria Lara and George Hasemann, extended hospitality and advice. I am grateful as well to John Warren, Peter Hearn, Delbert N. McCluskey, and several other individuals at USAID/Honduras who expedited the study in many different ways. Special assistance came from Bessy LaPaz, of the Honduran National Census, who made unpublished census data available to me whenever I wished. Over the years, the staff of the Panamerican Agricultural School at Zamorano, especially Dan Meckenstock, were very helpful in providing me with reports of their own research. In the last few years I have had the pleasure of working with Miriam Dagen and Roland Bunch of World Neighbors, who aided me in every way feasible. I also would like to acknowledge the help of Denise Stanley and Lyn Moreland, who conducted fieldwork related to the expansion of shrimp farms. Several individuals read and made useful comments on earlier versions of this text, including Billie DeWalt, Jeff Boyer, Sandy Robertson, and Doug Murray. Special thanks, as well, go to my husband, Jerry Sorensen, who not only read every word in the manuscript but, as chief editor, read each one a hundred times. Scott Whiteford and Bill Derman, series editors, and Kellie Masterson, senior editor at Westview Press, critically assessed the later versions. Several individuals from the Departments of Anthropology and Environmental Studies at the University of California at Santa Barbara also aided in the production process, especially Dirk Brandts, Mary Gervase, Judy Klinge, and Eric Meyer. It is my hope that all the above people and the many others who are concerned with ameliorating human impoverishment and environmental destruction in all parts of the world find this book useful.

Susan C. Stonich

Introduction

*I can only expect destruction for my family because I am provoking it with my own hands. ...
This is what happens when the peasant doesn't receive help from the government and the
banks.... He looks for the obvious way out which is to farm the mountain slopes and cut down
the mountain vegetation. Otherwise how are we going to survive? We're not in a financial
position to say, "Here I am! ... I would like a loan to plant so many hectares!" I put in my request but the banks don't want to give me credit because I cannot guarantee the loan. I know
what I am doing ... as a person I know ... I am destroying the land!*
                                                            —Southern Honduran peasant, 1990

Bounded on the southeast by Nicaragua and on the west by El Salvador,
southern Honduras was flanked by political revolutions during the 1980s.
While media coverage focused on military actions and human suffering in
those countries, the desperate and worsening circumstances of the people of
the south largely went unnoticed. Yet, according to most measures of income,
health, and nutrition, the situation of most southern Hondurans was as bad or
worse than that of their neighbors in El Salvador and Nicaragua. Located
within the Pacific watershed and characterized by steep slopes, rugged terrain, and erosive soils, the south is also the most populated and environmentally degraded area of Honduras. The underlying and more immediate causes
of the complex human and environmental problems there are the focus of this
book.

This work is about interconnections—among the historical, social, economic, ecological, and demographic aspects of development—and about the
ways Honduran people are struggling with increasing poverty and environmental destruction. Using an approach which integrates political economy
and ecology, the book examines the systemic linkages among the dynamics of
agricultural development, demographic change, associated patterns of capitalist accumulation, rural impoverishment, and environmental decline. It is a
specific historical example of how larger international and national forces
evolved and affected people and the natural environment and how, in turn,
human agency is affecting those powerful forces. In so doing, the study contributes to an understanding of the complex interrelationships between development and the environment, to predictions regarding the destiny of
peasantries within the dynamics of contemporary capitalism, and to the recognition that the fates of the peasantry and the natural environment are intimately linked. Moreover, the complex, multifaceted interconnections established for the Honduran case are widespread throughout much of Central
America.

This work also examines another link—the tie between Honduras and the United States—which to a great extent has shaped these other interconnections. Although all countries of the isthmus have been under the direct influence of the U.S., none has been more completely controlled by their economic, political, and military relations with the U.S. than has Honduras. From the beginning of this century, Honduras' principal economic activities have been dominated, directly or indirectly, by U.S. companies. After World War II, U.S. economic interests were bolstered by the assistance of the United States Agency for International Development (USAID) and other U.S. bilateral agencies—most recently evident in the promotion of so-called nontraditional exports. According to conservative estimates, in the mid-1980s, 60 percent of the Honduran economy was controlled by U.S. companies (Paz 1986: 190). While U.S. banana companies, Chiquita (formerly United Fruit and later United Brands), Castle and Cooke, and Del Monte (R. J. Reynolds) have dominated the Honduran economy, a study by the National University's Institute of Economic and Social Research found that U.S. transnational corporations own 100 percent of the country's five largest firms, 88 percent of the 20 largest, and 82 percent of the 50 largest (Barry and Preusch 1986). Moreover, the U.S. is the principal buyer of Honduran exports, the main supplier of Honduran imports, and the biggest creditor for Honduran external debt (Paz 1986: 190).

Honduras, however, played a marginal role in U.S. foreign policy until after the Sandinista Revolution in 1979 when its strategic position was reevaluated. In exchange for massive economic and military aid during the 1980s, Honduran officials submitted to the demands of the Reagan administration and militarization of the country accelerated. Paradoxically, U.S. policy toward Honduras was not implemented on behalf of Honduran interests but rather on behalf of Washington's goals to destabilize the Nicaraguan government and to defeat, militarily and politically, the Salvadoran guerrillas (Shepherd 1985). In part, this work examines the human and environmental consequences of thrusting Honduras into the framework of U.S. political strategy. It does not focus directly on the results of militarization but rather on the effects of the economic development initiatives that accompanied U.S. intervention—efforts based on a dominant development model that persists into the 1990s despite overwhelming evidence of its excessive human and environmental costs.

During the 1980s, the deteriorating social, economic, and ecological conditions in all of Honduras, including the southern region, were eclipsed by more dramatic events in neighboring countries and were ameliorated by enormous amounts of foreign assistance. Honduras confronted the 1990s under difficult circumstances: the implementation of a stern structural adjustment program; the fourth lowest Gross Domestic Product (GDP) per capita in Latin America—estimated at US$483 for 1990 (higher only than that of Bolivia, Haiti, and Nicaragua); excessive poverty—70 percent of the total population and 80 percent of the rural population live below the absolute poverty level; and an extremely skewed income distribution in which 40 percent of the poorest receive 7.3 percent of the total income while the richest 10 percent collect 50 percent (CONAMA 1991: 46, 57–59). Nationally, undernutrition affects

40 percent of school-age children (SAEH/INCAP 1987) and in some areas, 70 percent of preschool children suffer from malnutrition, of which 30 percent reaches second- or third-degree levels (McBride 1992). Honduras ranks last among Central American countries according to the relatively new estimate of "progress" created by the United Nations in 1990, the Human Development Index (HDI)—a composite measure calculated from multiple indicators of life expectancy, educational attainment, and income (UNDP 1992).[1]

In 1990, the population was estimated at 5.1 million people nearly double the 1970 population of 2.6 million (World Bank 1992: 268). Moreover, while the total fertility rate dropped from 7.4 in 1970 to 5.3 in 1989 and the annual growth rate declined from 3.7 percent in 1980 to 2.96 percent in 1990, more than 46 percent of the population currently is less than 15 years old—that is, just entering into childbearing years—and the country's population will continue to grow significantly into the next century (World Bank 1992).

High population growth rates and lack of economic alternatives have forced growing numbers of rural peasants to put unprecedented pressure on natural resources by adopting destructive agricultural practices in their home areas or by seeking survival elsewhere, either in the rapidly expanding urban centers or in the last vestiges of tropical humid forest. The average annual rate of urbanization remains at approximately 5.5 percent—the highest in all of Central America (World Bank 1992: 264–265). Estimates of the current rate of deforestation range from 1.6 percent to 3.2 percent annually or about 80,000–90,000 hectares (CONAMA 1991: 14; WRI 1992: 286) and soils are being lost at the rate of 10,000 hectares per year (USAID 1990: 3).

Central America in the 1990s

With media attention and public concern focused on Europe and the Middle East there is a feeling that relative calm exists in Central America. But the peace is a delusion and long-term conflicts continue throughout the region. Although peace may have been proclaimed by the presidents of the Central American republics, what cannot be decreed are that the interrelated social, economic, and environmental conditions which aroused class struggle and armed intervention have been resolved. While Central American presidents issue declarations regarding the achievement of stable democratic societies, sustainable economic growth, and effective regional cooperation, the region remains mired in long-term, interrelated, and mutually reinforcing crises. The economic crisis of the 1980s during which Central American governments struggled with the burden of servicing external debts in a context of deteriorating terms of trade has extended into the 1990s. Recently elected Central American governments committed to market-based economic policies have responded to overwhelming pressure from international donor and lending institutions and have established severe economic adjustment programs. These measures have resulted in declines in real income and social services especially among the most vulnerable segments of society—the rural and urban poor. As a result, poverty accompanied by diminished nutrition and health is expanding and social discord continues to foster political unrest in every country of the region.

In addition to these profound problems is a related crisis—an ecological crisis characterized by rampant environmental destruction. Abundant evidence indicates that the rates at which forests, soils, fisheries, and other critical natural resources are being depleted far surpass the pace at which they are being renewed and that ensuing effects such as land degradation, watershed deterioration, and destruction of coastal resources have reached acute levels in many areas (Leonard 1987; USAID 1989c; Faber 1992b). The extent of this crisis is revealed dramatically in the destruction of the region's tropical forests. Rates of forest clearing rose significantly after 1950 and by the early 1980s, less than 40 percent of Central America's forests remained standing (Leonard 1987: 117). The World Resources Institute estimated that during the 1980s, 400,000 hectares of Central American forests were lost annually and that the total deforested area increased by approximately 50 percent (WRI 1992: 118–119). During the decade, the annual average rate of deforestation in Central America was 2.3 percent—higher than the rates in either Brazil (0.5 percent) or Mexico (1.3 percent). Among Central American countries, average yearly rates ranged from 0.6 percent in Belize, to 0.9 percent in Panama, 2.0 percent in Guatemala, 2.3 percent in Honduras, 2.7 percent in Nicaragua, 3.2 percent in El Salvador, and 3.6 percent in Costa Rica (WRI 1992: 286).

Furthermore, by 1992, 24 percent of the land in Central America exhibited moderate to extreme soil degradation—the highest percentage of significantly degraded land in the entire world (WRI 1992: 116). Most soil degradation (77 percent) is in the form of water and wind erosion and is found within the mountainous Pacific watershed, the result of agricultural activities, overgrazing, deforestation, and fuelwood collection. Chemical deterioration accounts for about 11 percent of soil degradation: primarily due to loss of soil nutrients (from deforestation and unsustainable agricultural practices), salinization (primarily the result of poorly drained irrigation systems and the excessive loss of groundwater), and pollution (associated with the indiscriminate use of pesticides and the uncontrolled expansion of urban areas). Compaction caused by cattle contributes to the physical deterioration of soils as well (WRI 1992: 116).

The human dimensions of the ecological crisis also are profound. The critical social and economic changes that have occurred in the region point to the intimate relationship between environmental conditions and human health. As a result of inadequate sanitation measures, diarrheal and parasitic diseases are a leading cause of high mortality and morbidity among certain population groups—especially among the poor, women, and children (Jacobson 1992). Although data linking environmental deterioration and health in Central America are incomplete, sufficient data do exist to suggest that the population's health has been adversely affected by high levels of exposure to chemical substances and environmental contamination (PAHO 1990).

Continuing urban growth has contributed to elevated health risks related to increased exposure to toxic chemicals contaminating the air, water, soil, and food (PAHO 1990). Between 1965 and 1990 the percentage of Central Americans living in urban areas rose from 37 percent to 48 percent and the urban population continues to increase by approximately 3.6 percent annually (World Bank 1992). Deteriorating living conditions are a typical aspect of large cities in the region, especially in newer squatter settlements and in decaying downtown areas. Families that migrate to the cities from rural areas often are unable to find adequate housing as well as employment—aggravating a housing problem made more acute by the persistent economic crisis. Poor living conditions, including scarcity of potable water and sanitation, are getting worse as population densities increase and public services (including health) continue to decline (ECLAC 1990; PAHO 1990).

In addition, efforts to cope with economic crisis have led to the rapid emergence and growth of industries whose effects on health and the environment have not been assessed sufficiently (Faber 1992a; Goldrich and Carruthers 1992). The production and use of chemicals employed in such industries located within the growing number of free trade zones, significantly increase the risk of environmental health problems (Ives 1985). At the same time, increased accidents, noise, and congestion also adversely affect environmental health. Because the region as a whole is committed to economic growth, the upward trend in environmental problems affecting human health will persist and is likely to accelerate (PAHO 1990).

The link between environmental decline and human health in rural areas and the evidence for growing rural poverty indicate that rural areas, too, are increasingly at risk. Rural people have even more diminished access to a safe and adequate water supply, excreta disposal, improved housing, and health services. Furthermore, because the rural poor are increasingly dependent on off-farm income, much of which is earned as agricultural laborers, they also are at an elevated risk regarding proper procedures for handling and using pesticides (WHO 1990). Recent work in Mexico (Wright 1986, 1990) as well as in Central America (Leonard 1987; Swezey et al. 1986; Thrupp 1988, 1991; Murray and Hoppin 1990, 1992; Murray 1991) indicates that the excessive use of pesticides is placing farm workers in jeopardy as well as contaminating the soils, surface, and groundwater thus endangering long-term sustainability. In addition, as some vectors have become resistant to pesticides, the incidence of diseases once thought to be on the decline, such as malaria, is rising.

All Central American countries have been undergoing a severe economic crisis since the beginning of the 1980s, and in some of the countries major natural disasters, political unrest and wars have exacerbated the effects of the crisis. As a result of these and other factors, the overall rise in food prices outstripped increases in wages, resulting in a drop in real purchasing power (ECLAC 1989). Because food consumption is linked to real income levels, these data suggest that food consumption levels most probably deteriorated as well—especially among the poor. In addition, data on food production show that per capita food production declined an average of 15 percent among all countries in the region between 1980 and 1990 (World Bank 1992).

Overall information on national and per capita food production, however, reveals only part of the reality. A recent analysis of hunger and food security in Central America (Corbett 1991; Whiteford and Ferguson 1991) demonstrates how hunger and malnutrition can exist simultaneously with growth in agricultural production, especially if crops are destined for foreign markets. The authors show how international and national class interests and power relationships have meshed with donor and debt strategies to create food and nutritional deficits in many Central American communities. They point to several factors which have generated food insecurity throughout much of the region: extreme socioeconomic and political inequality; war and the large-scale displacement of peoples; the international debt crisis; overemphasis on export agriculture and ineffective support in the form of credit, services, and technologies for food-crop producers; over-reliance on food imports; poorly developed mechanisms for distributing food to those most in need; and the depletion of the natural resource base emanating from development strategies. The debt crisis, population growth, and the desperate need of rural people to sustain their families have all led to efforts to increase foreign exchange earnings. These efforts in turn have led to the staggering rate of environmental destruction. The expansion of export crops and the inequality in land tenure in Central America have forced thousands of smallholders off the land and contributed to growing landlessness in rural areas. Those who remain often find themselves relegated to more marginal, steeply sloped lands that are easily degraded. The severe soil erosion and other forms of environmental degradation have decreased crop yields at a time when peasants face severe economic hardships. This process of marginalization of peasant producers has generated hunger, reduced national food production, and systematically destroyed the resource base. The health of the natural environment has become a critical dimension of present and future food insecurity in the region.

Conserving Central America's natural resource stocks is crucial because the region continues to depend on natural resources to generate income in such key economic sectors as agriculture, forestry, and fisheries. Natural resource based commodities also persist as the principal means of earning foreign exchange: in 1990, the average percentage of total export earnings from such commodities was 81 percent: ranging from 72 percent in Costa Rica, to 74 percent in El Salvador and in Guatemala, 78 percent in Panama, 85 percent in Honduras, and 94 percent in Nicaragua (World Bank 1992). Despite significant economic dependence on natural resources, Central American countries continue to destroy resource stocks in the pursuit of foreign exchange with little regard for the long-term environmental costs. In calling for new ways to include environmental costs into economic measures of "progress" or growth, Robert Repetto could have been referring specifically to Central America when he said, "A country can cut down its forests, erode its soils, pollute its aquifers, and hunt its wildlife and fisheries to extinction, but its measured income is not affected as these assets disappear. Impoverishment is taken for progress." (Repetto 1992: 94).

The chronic poverty, food insecurity, and environmental degradation in Central America cannot be understood apart from the dominant development strategy which Central American governments have implemented. While structural adjustment policies in the areas of exchange rates, pricing, marketing, and public sector reform may be crucial elements of national economic strategies, the prevailing development policies promoted by bilateral institutions such as USAID and multilateral institutions including the World Bank, the International Monetary Fund (IMF), the Inter-American Development Bank (IDB), and others have had severe social, economic, and ecological costs. Short-term policies, designed to increase commodity exports to generate foreign exchange, often clash with longer-term development needs. Too often, scarce land, water, credit, and technology are channelled to the export sector while the requirements of small farmers are disregarded. Land reform, credit, improved extension, greater government accountability, and peoples' participation are important prerequisites if the conditions of the growing number of landless and landpoor peasants are going to improve. Even in Costa Rica—at one time considered the Central American success story despite having excessive levels of per capita external debt and one of the highest deforestation rates in the world—large numbers of people failed to share in the benefits of the economic growth of the 1960s and 1970s (Seligson 1980). As governments throughout the isthmus are compelled to cut public expenditures and restrict imports as part of IMF or World Bank economic restructuring programs, the lives of growing numbers of Central Americans deteriorate further. In Costa Rica, the government's economic reform measures instituted between 1989 and 1991 led to a reduction of the fiscal deficit, cut the trade deficit significantly, and boosted foreign reserves. At the same time, however, economic growth slowed from 5.9 percent to 1 percent due to worsening terms of trade for Costa Rica's traditional exports (especially coffee) and to the recession in the developed world. Critics of government imposed structural adjustments maintain that the reduction in Costa Rica's fiscal deficit is due largely to the elimination of thousands of public sector jobs and the curb on public-sector wages. Combined with the partial lifting of price controls and rising inflation which reached 31.2 percent between April 1991–1992, these factors contributed to a decline of 15 percent in the real wages of public sector workers and 10 percent in the wages of private sector employees in 1991. Between 1987 and 1992, state spending on education, health, and other programs aimed at the poorer segments of Costa Rican society fell by 30 percent and during the first two years of President Calderon's administration, the number of Costa Ricans living in poverty rose from 400,000 to 600,000 (LARR 1992c).

Although much attention has been paid to the economic and financial problems that have plagued the region, these disasters are intensified by the environmental crisis, to a great degree caused by unsustainable, elite-oriented development strategies. The environmental crisis in turn exacerbates the economic crisis as worsening land degradation contributes to declines in food production while also advancing Central America's dependence on food aid and imports. Augmenting the problem is population growth. Although the average total fertility rate for the region dropped significantly from 6.7 children to 4.5 children between 1965 and 1990, more than 40 percent of Central America's population is less than 15 years old and the average annual growth rate of 2.6 percent remains among the highest in the world (World Bank 1992). Thus mounting social, economic, and environmental problems are taking place in a context of expanding populations thereby making potential solutions even more complex and continuing the downward spiral in human and environmental conditions.

Agricultural Diversification and Export-led Growth

Central American governments have a long history of pursuing agricultural policies that support the expansion of export crops and livestock production (Williams 1986). Williams (1986, 1991) argues that the recent political crisis in Central America can be traced to processes that evolved from basic structural inequalities, the enclosure of traditional peasant lands, and the modernization of agricultural labor relations resulting from this export focus. Liberal economic initiatives implemented by many Central American governments during the nineteenth century to expand the production of export crops, particularly coffee, resulted in the privatization of many communally owned and managed lands. These and other actions to promote exports led to the concentration of property in the hands of a small minority and to the spreading marginalization and impoverishment of the rural population. For many families, diminished access to land led to greater dependence on wages earned as seasonal laborers on large farms (Williams 1986, 1991).

Development schemes aimed at alleviating Central America's social and economic problems historically have stressed intensified exploitation of the region's natural resources through augmented exports of agricultural commodities and forest products, enhanced agricultural productivity, and expanded industrial fisheries. Export growth through agricultural diversification remained the cornerstone of development strategies in Central America throughout the post–World War II period. In response to increasing political unrest in the region that were perceived as threats to economic and national security, U.S. policy initiatives emphasized economic development in combination with military assistance. In the aftermath of the Cuban revolution, President John Kennedy began the Alliance for Progress in 1961: a multibillion dollar development effort founded on export diversification and the expansion of agricultural export earnings—especially cotton, sugar and beef—the "nontraditional" agricultural exports of that era. International assistance and donor agencies such as USAID and the World Bank promoted efforts to stimulate trade, to modernize production, and to establish credit programs to help expand the production of nontraditional commodities. These programs were directed into the region through the state and the augmented investment of multinational corporations and national elites (Stonich and DeWalt 1989).

At first glance, the Alliance for Progress seemed effective: until the early 1970s the economies of the region grew at 6 percent annually, average per capita income rose by more than 2 percent a year, and export earnings grew by approximately 10 percent annually (Williams 1986). Central America's dependence on traditional exports of coffee and bananas decreased as well: from 60 percent of export earnings in 1961 to 40 percent in 1973 (Williams 1986). However, these economic gains improved the standard of living of only a small proportion of the population as land ownership became more concentrated and control of production became dominated by large-scale national and international investment. The inequitable distribution of wealth in the region was exacerbated as a relatively small elite benefitted disproportionately from the economic growth of the period (Bulmer-Thomas 1987).

During the 1970s, due to declines in international commodity prices, soaring costs for oil and technological inputs such as pesticides, the collapse of regional markets, and the growing debt burden, the region sunk into severe economic crisis distinguished by a decline in economic growth and by worsening inequality (ECLAC 1986; Bulmer-Thomas 1987). By 1980, 63.7 percent of Central Americans (76.2 percent in rural areas) were living in poverty without sufficient income to cover essential needs such as food, housing, clothing, and basic services (ECLAC 1986: 20). The impact of the economic crisis was particularly severe on the poor, and dissension and repression grew in tandem with the intensified economic crisis (Williams 1986; Brockett 1988). By the end of the 1970s, bolstered by significant urban labor, middle class, and student support, peasant uprisings and resistance accelerated, culminating in revolution in Nicaragua, ongoing civil wars in El Salvador and Guatemala, and the militarization of Honduras.

In response to deteriorating social and economic conditions, U.S. policy makers in the early 1980s once again proposed economic reforms in order to defuse the political crisis in the region. The strategy was implemented in 1984 through the Caribbean Basin Initiative. The goal was to establish regional stability through economic growth stimulated by domestic and foreign investment. Agricultural diversification through the promotion of a new set of nontraditional exports was the foundation of this scheme. The most important of these nontraditionals included: fresh, frozen, processed and otherwise preserved fruits and vegetables (e.g., melons, snow peas, broccoli, eggplant; root crops; edible nuts; live plants and cut flowers; and the most commercially desirable species of crustaceans and mollusc—especially shrimp and lobster (Paus 1988). The objectives were to broaden export earnings to stimulate domestic growth while facilitating the servicing of the region's foreign debt and to increase income among the region's small producers thereby encouraging more equitable growth and averting further social unrest (Murray and Hoppin 1992). Central American governments currently are increasing their emphasis on the nontraditional agricultural export sector and seeking ways to diversify their export base to generate larger amounts of foreign exchange. Foreign exchange earnings from nontraditional exports rose from US$423 million in 1983 (approximately 12 percent of total export earnings from the region) to US$1.3 billion in 1990 and have the potential of reaching US$4 billion by 1996—about 50 percent of total export earnings (USAID 1991b).

The new nontraditional commodities are championed not only as new sources of foreign exchange but also as a solution for widespread rural poverty. In contrast to traditional export crops, they are promoted on the basis that they lend themselves to small scale, labor intensive production by small farmers and bring a higher return than traditional basic grains thereby contributing to increased rural income and more equitable distribution of wealth. According to Anthony Cauterucci, who was USAID mission director to Honduras at the time: "We are looking at moving subsistence-oriented agriculture toward high commercial value agriculture and linking *campesinos* to the export process." (Garst and Barry 1990: 88). While it may be that these new crops poten-tially are more profitable than basic grains, their widespread expansion raises serious distributional questions. To be profitable, nontraditional crops require significantly higher investments in a variety of expensive inputs (seeds, fertil-izer, pesticides, irrigation) and greater access to technical expertise and inter-national markets than do basic grains. Furthermore, as nontraditional farming expands, land values may increase, squeezing smaller producers off their holdings—augmenting the already high number of landless farmers in the re-gion. Without adequate redistribution measures, nontraditional agro-export promotion may lead to an even more inequitable distribution of income and further impoverish rural populations. (Garst and Barry 1990: 89–90). In addi-tion, nontraditional crops bear considerable risk for small farmers. They must be grown to increasingly rigid standards and harvested at specified times within a context of fierce global competition. If, for any reason, the crop is re-jected, small producers generally have little access to other markets. Neither can farm families subsist on unsold flowers and so-called Oriental vegetables. Moreover, global agricultural commodity markets are notoriously unpredict-able, as Central American producers of more traditional commodities (such as cotton, sugar, bananas, and coffee) and of more recent nontraditonals (includ-ing sesame, cardamon, and frozen beef), already are aware. Finally, protec-tionist policies implemented at the insistence of recipient country producers can reduce imports independent of demand—as happened in the instances of U.S. beef and flower producers during the 1980s.

The growing body of evidence from studies that directly address issues re-lated to the expansion of nontraditional exports in Central America suggests that their impact on the rural poor and the natural environment may not be fa-vorable (e.g., Krueger 1989; von Braun et al. 1989; Garst and Barry 1990; Mur-ray and Hoppin 1990, 1992; Murray 1991; Rosset 1991; Stonich 1991c, 1992). These studies note striking parallels in the social processes associated with the recent boom in nontraditionals and the earlier expansions of other natural re-source based export commodities (coffee, cotton, sugar, and livestock). They suggest that the promotion of nontraditional exports may be recreating or ex-acerbating the social and ecological crises associated with commodities pro-moted during previous periods. Serious concerns have emerged regarding eq-uity, the actual increases in household income generation, the increased marginalization of small producers, the disadvantageous position of women workers in the nontraditional export processing plants, the effects on diet and nutrition, the excessive use of pesticides, and the further destruction of the natural resource base. The indications are that similar to previous export booms, the promotion of nontraditionals may create a variety of economies of scale that accelerate social differentiation in rural areas, expel large numbers of small peasant producers from their lands, and lead to intensified social in-stability and conflict (Rosset 1991).

The Role of Foreign Assistance

Central American governments are subsidized in their efforts to promote export-led growth by many bilateral and multilateral, international donors and lending institutions. During the 1980s, the region experienced a tremendous growth in so-called non-military, economic assistance from these agencies. The majority of bilateral assistance came from the U.S. as part of its efforts to stabilize regional governments and to appease populations in order to prevent successful popular uprisings such as the Sandinista victory in Nicaragua in 1979. Although various U.S. organizations provide bilateral assistance to the region, the most significant is disbursed by USAID. Between 1946 and 1979, USAID dispensed a total of US$1.9 billion to Central America; between 1980 and 1990 alone, it disbursed US$7.5 billion (These and the following years and aid levels are calculated on the basis of government fiscal years and data from *USAID, Congressional Presentation*, various years). During the second half of the decade, U.S economic aid made up an average of 62.4 percent of total official development assistance (ODA) to the region (excluding Nicaragua which did not receive financial assistance from the U.S. between 1984 and 1989).

USAID administers three major types of economic assistance programs: Development Assistance (DA), Public Law 480 (PL 480) food aid, and Economic Support Funds (ESF).[2] Economic Support Funds to major recipients—Costa Rica, El Salvador, Guatemala, and Honduras—grew from US$9 million in 1980 to a maximum of US$502 million in 1984 before declining to US$220 million in 1992. Between 1980 and 1990, ESF averaged US$300.5 million annually or 50.5 percent of all non-military economic assistance to the region. Under Section 502-B of the 1974 Foreign Assistance Act, ESF are classified as part of the Security Assistance Program which also includes Foreign Military Sales, the Military Assistance Program, and the International Military Education and Training Program. The majority of ESF is disbursed as cash transfers from the U.S. treasury to central banks of recipient governments that are politically strategic and friendly to the U.S. Although U.S. law forbids this kind of balance-of-payments support from being used directly for military purposes, ESF funds free up other money to be used for such purposes (Danaher et al. 1987). The practice of classifying ESF as economic aid incorrectly leads many to conclude that the majority of U.S. aid to Central America during the 1980s was used for economic development activities. In reality, DA averaged only US$135.3 million annually during the decade—about 30.7 percent of total economic assistance.

The majority of food aid to Central America in the form of PL 480 functions similarly to ESF: i.e., as budgetary transfers which stabilize governments, pacify people, and free up the resources of recipient governments to be used for other things (Garst and Barry 1990). Throughout the 1980s, PL 480 operated through two major programs: Title I (a concessional sales program) and Title II (donations for free distribution). A third program, Title III was set up in 1978 specifically for the lowest-income countries. Title III forgives Title I dollar repayments for the poorest countries that agree to implement specified policy reforms and development projects. Only Honduras received Title III support during the 1980s because it was the only Central American country with a per capita income low enough to qualify. Through Title I, governments receive easy credit (loans of up to 40 years at 2–4 percent interest) to import U.S. agricultural commodities which are then sold on the open market (Cloud 1990: 2767). Between 1980 and 1990, US$772 million was channeled to Central America through Title I/III program assistance—about 70 percent of total U.S. food aid during the period.

Public Law 480 programs also are used to support the expansion of nontraditional commodities. Local currency available through Title I is used to augment recipient government incentives for the new crops including credit for private agribusiness, extension, marketing, and processing (Garst and Barry 1990: Chapter 2). An evaluation of the PL 480 Title I program in Honduras estimated that less than one percent of Title I local currency was designated for the production of basic food grains (Norton and Benito 1987). At the same time, food rations available through Title II agreements maintain small farmers and colonizers while they gamble on establishing the new nontraditionals (Garst and Barry 1990: 88).

There are several serious criticisms of PL 480 food aid programs in the region. Among them are that the development and nutritional aspects of the programs largely have been usurped by political interests. Further, because most food is sold rather than distributed without cost, the most susceptible segment of the population, the poor, have limited access. Finally, food aid serves as a disincentive for national producers who cannot compete with the subsidized prices (Danaher et al. 1987; Garst and Barry 1990; Whiteford and Ferguson 1991). Between 1975 and 1990, the amount of food aid by weight to all Central American countries swelled by 2,000 percent and the quantity of cereal imports grew by 187 percent while during approximately the same time period (1979–1990) the average index of food production fell to 85 percent of 1979 levels (World Bank 1992). Evidence suggests that among the effects of augmented food aid has been a growing dependence on food imports and little indication that PL 480 programs have reduced the widespread hunger in the region (Garst and Barry 1990: 89; Whiteford and Ferguson 1991).

According to USAID's, "new" strategy for U.S. economic assistance for Central America in the 1990s, donor assistance from multilateral institutions is expected to be substantially larger than during the 1980s (USAID 1991b). While net disbursements during 1984–1990 from the IMF were negative US$400 million, for 1991–1996, net flows with the IMF are estimated to be about zero because the region has substantially reduced its indebtedness to the Fund. As long as Central American countries continue to comply with the demands of the World Bank/IMF and IDB, disbursements of funds from those institutions should more than double during the 1991–96 compared to the previous six years. USAID also predicts a modest increase in other bilateral donor flows during the 1990s which should allow a gradual decrease in U.S. assistance to the region. Under the scenario presented by USAID, assistance would decline from US$810 million in 1990 (excluding US$420 million in extraordinary assistance to Panama) to US$430 million by 1996 for a total of US$3.2 billion between 1991–1996. This would be a decline from US$6.2 billion disbursed between 1984–1990. Over US$1 billion of assistance during 1991–1996 is expected to flow to Nicaragua. For the four countries included in the Central American Initiative during 1984–1989 (El Salvador, Guatemala, Honduras, and Costa Rica) the decline would be from US$6 billion in 1984–90 to US$2 billion during 1991–1996 (USAID 1991b). Thus, through the 1990s, Central American countries will likely have to cope with reduced development assistance and enhanced indebtedness to the World Bank and the IMF.

Conclusion

In view of the overwhelming evidence of intensified environmental destruction and human impoverishment taking place throughout Central America, it is essential that development efforts take into account the interrelationships between the economic changes that are introduced and their human and environmental repercussions. This book examines the consequences of the prevailing development model on one region within Central America—southern Honduras located within the Pacific watershed along the Gulf of Fonseca—where the growth of agricultural exports has been far-reaching and which is characterized by significant demographic changes, expanding impoverishment, food and resource scarcity, and environmental decline. Analysis centers on the most important exports that have been promoted in the post–World War II period: cotton, cattle, and the most recent array of nontraditional exports, especially cultivated shrimp and irrigated melons.

This book begins with the premise that the persistent crises in Central America are not due primarily to the actions of escalating numbers of backward peasants unaware of the serious repercussions of their acts. While mounting populations, war, and the worst drought in decades are partly to blame, much of the poverty and environmental destruction is the result of misguided national development policies often financed by international donor agencies and lending institutions. The ecological crisis in Central America cannot be understood in isolation from the structural processes that are determining the way natural resources are used. Policy reforms such as devaluation, pricing, and budget and tax reforms are not enough to ensure sustainable agricultural production, sound management of natural resources, and reductions in poverty and inequality.

It is becoming increasingly apparent that the economic achievements of the 1960s and 1970s were illusionary in the long-term and that they occurred at a major cost to the present and future's capacity to extend economic growth—and even at the more serious price of an actual decline in human welfare. For the most part, the persisting emphasis on export-led agriculture ignores the resource requirements of the growing millions of landpoor farmers and landless families. Because scarce land, water, credit, and technology are being preempted by the export sector, poor farmers, lacking financial means and technical support, over-exploit the limited natural resources they control in order to eke out a living. Outcomes have included widespread environmental destruction and the reduced capacity of Central America to feed itself.

The following chapters demonstrate the interrelationships among development, poverty, and the environment in southern Honduras. Chapter 1 begins with a discussion of the potentially complementary intellectual trends within and outside of anthropology that have taken place recently and have provided the stimulus and foundation for this work. These include the growing recognition of the interrelationships among development, population, society, and the environment; theoretical and methodological advances within anthropology related to expanding the traditional locus of anthropology beyond the community; and the status of the debate concerning the destiny of Latin American peasants. The remainder of the chapter takes issue with the failure of dominant paradigms to explain adequately environmental destruction and impoverishment in tropical areas of the developing world and provides an alternative approach that integrates political-economic and human ecological analysis. Subsequent chapters introduce, analyze, and integrate the various factors and levels of analysis that link development efforts with significant demographic dynamics, social processes, and environmental destruction. Analysis begins at the regional level in Chapter 2 with an examination of the several sets of linkages operating between population and the environment. Special attention is given to the interrelationships among the spatial distribution of natural resources and people, the most serious problems of environmental destruction in the region, and crucial population dynamics—especially growth, migration, and urbanization. The historic dimension, especially the importance of history to present circumstances is the heart of Chapter 3. The brief socioeconomic history of southern Honduras, from prehistory to World War II, demonstrates how specific political-economic linkages evolved in the region and facilitated subsequent development efforts. Although complete integration of southern Honduras into the world economic system occurred after World War II, Chapter 3 shows that many of the demographic, social, and ecological conditions that existed at the time of more complete integration were derived in no small measure from earlier periods of partial integration.

Chapter 4 also focuses on the region and chronicles the transformation of southern Honduran agriculture in the context of the global capitalist expansion that began after World War II. It describes the macroeconomic contexts that affected the decisions of the Honduran government and the subsequent integration of southern Honduras into the world economy through international and national efforts aimed at increasing and diversifying agricultural exports. This chapter emphasizes the effects of this development strategy on the allocation and distribution of land, on food production, on the overall availability of economic resources (employment and income distribution), on regional demography, and on human nutrition and well-being.

Chapter 5 considers the effects of the prevailing development model on more micro-levels of analysis—municipalities and rural communities. Based in part on ethnographic and survey research conducted between 1981 and 1991, this chapter attempts to determine commonality and variation in local level responses to the constraints and incentives posed by the regional processes already discussed. Although integrating the results of research conducted in a sample of nine highland, foothill, and lowland communities, the chapter centers on changing conditions during the 1980s in two adjacent highland communities—San Esteban and Oroquina. Although the two communities share similar environmental and natural resource contexts, they differ in significant ways in terms of the distribution of those resources, the degree and nature of social and economic differentiation, in overall patterns of the types of households and in the household economies, the existence of networks within and beyond the community, the importance of off-farm income, the extent and patterns of migration, and the degree to which community members are able to meet their nutritional needs. They diverge, as well, in their abilities to cope with the worsening economic crisis of the 1980s. While the focus of Chapter 5 is on community patterns, Chapter 6 centers on rural highland households as the major units of production and reproduction. Of particular concern are the ways in which larger forces stimulated the processes of socio-economic differentiation and proletarianization and affected household economic survival strategies. Diversity in the household economy is emphasized through an examination of multiple livelihood strategies which combine work on family based farms, wage labor, and service jobs within and beyond local communities. Also of great concern are the importance of monetized income in general and of migration in particular as a critical household income generating activity and as a link to broader regional, national, and international processes. In the second part of the chapter, the major causes of deforestation and land degradation in highland and foothill zones are discussed and found to be linked to the prevailing development model. The interaction between the agricultural and natural resource practices of various socially differentiated groups (e.g., resource-poor small farmers and renters, medium farmers, large farmers, and corporate farmers) are shown to exacerbate and mutually reinforce the overwhelming pattern of environmental destruction in the region. Also revealed is the potential conflict between ecologically sustainable agricultural practices and the household survival strategies of the rural poor.

The conclusions in Chapter 7 emphasize the integration of the various demographic, social, economic, and environmental crises within the region, the implications of current trends in development especially concerning issues of equity and environment; and the relationship between the southern Honduran case and the rest of Central America. The overall implications of the study on understanding the linked destinies of peasants and the natural environment also are discussed. Finally, several policy alternatives based on these conclusions and aimed at reducing impoverishment and environmental destruction in the region are suggested: the extensive use of applied political-ecological analysis; increased access to productive resources, including land and opportunities for off-farm employment, despite the recent passage of the "agricultural modernization" law which, in effect, closed the book on the agrarian reform initiated in 1962; an approach to development which is based on a diversified rural society and which considers the value of women's unpaid domestic labor and the articulation between economic production and reproduction in peasant households; and the promotion of local organizations and initiatives through which communities can plan their own futures with dignity, generate income, and expand social services.

Notes

1. The algorithm used to calculate the HDI can be found on pages 91–96 of the United Nations Development Programme, *Human Development Report: 1991*, 1992. The

HDI can range from 1.0 (the highest or best score in terms of human development) to 0.0 (the lowest or worst). In 1992, HDI values for the 160 countries included in the report ranged from .982 (Canada) to 0.052 (Guinea). Scores for the Central American countries were: Honduras .473; Guatemala .485; Nicaragua .496; El Salvador .498; Belize .665; Panama .772; and Costa Rica .842 (UNDP 1992: 127–129).

2. It is difficult to say who controls PL 480. The United States Department of Agriculture (USDA) purchases the commodities, arranges financing, and occasionally transportation, while USAID manages them after they leave the dock. Power to determine which countries receive food aid is vested in the Development Coordination Committee (DCC) composed of officials from USDA, USAID, the State Department, the Treasury Department, the Office of Management and Budget, and the National Security Council. The last several years has seen considerable infighting among these agencies and the office of the President and has led to demands for extensive PL 480 reforms (Cloud 1990).

1

Linking Development, Population, and the Environment: Perspectives and Methods

In June 1992, representatives from 178 nations converged on Rio de Janeiro for the historic United Nations Conference on Environment and Development (also known as UNCED or the Earth Summit), while some 2,000 non-governmental organizations (NGOs) met at the unprecedented concurrent event, the Global Forum. The Rio summits marked the twentieth anniversary of the United Nations Conference on the Human Environment which convened in Stockholm to consider how human activities alter the global environment. In 1972, there was little recognition that economic development contributes to environmental decline and even less empirical research focused on the connections. Twenty years later, the environmental costs of unrestrained economic growth and the simultaneous need for increased economic opportunity persist throughout the world. Over the decades, however, a growing acceptance of the intimate connections between development processes and the state of the natural environment also arose. The conclusions and recommendations of international councils, such as the United Nations World Commission on Environment and Development (UNWCED 1987), and the environmental initiatives on the part of major funding institutions, which occurred during that time, point to the expanding recognition that complex interconnections exist among economic, social, demographic, as well as environmental processes, related to development.

In the aftermath of Rio 1992, however, many questions emerged about the value of the meetings and the limited nature of the agreements that were reached (Lowe 1992). For example, issues related to population growth, critical links between development and the environment, were touched on only superficially—despite engendering rancorous debate and the warning of Norwegian prime minister, Gro Harlem Bruntland, in her opening address to UNCED that, "poverty, environment, and population can no longer be dealt with, or even thought of, as separate issues" (Holloway 1992). Representatives to UNCED from Third World countries declared that they would not discuss efforts to reduce population unless developed countries were willing to consider endeavors to decrease consumption. At the same time, participants at the Global Forum, perceived discussion aimed at controlling population as an infringement on women's rights and as a means of deflecting efforts away from eradicating the real causes of poverty and environmental destruction in the Third World (Holloway 1992).

While it is premature to consider the ultimate legacy of the Earth Summit and the Global Forum, their greatest feat may rest not in the insufficient commitments made by governments and organizations but in the way that the events shaped the future global agenda—by bringing out new international values of equity and environment and tying them to relations between rich and poor nations (Speth 1992). The bitter debates over population concerns may have some positive results as well, as environmental organizations increasingly include population issues into their endeavors, and foundations establish more programs aimed at integrating the population dimension into research agendas concerned with linking development and the environment (Holloway 1992).

Despite the growing recognition among policy makers and the public that complex links exist among development, population, and the environment, current research and practice in each area largely remain separate and professionals in one domain are only starting to comprehend the issues and priorities of the others (Brandon and Brandon 1992: 477). In addition, understanding the interrelated human and environmental problems currently being faced by the world, also demands the facility to shift among local, regional, and global levels. Since the 1970s, theoretical and methodological changes have occurred within anthropology which have strengthened anthropologists' abilities to address conceptual issues and methodological approaches that integrate development and the environment more fully both in terms of linking relevant domains of knowledge and levels of analysis. These advancements also have enhanced the participation of anthropologists in interdisciplinary research related to global environmental change.[1]

In response to criticisms that community focused anthropological studies did not adequately consider the relationships between such communities and the larger political and economic systems of which they were a part, many anthropologists expanded their research focus to examine how relatively small groups are integrated into larger regional, national, and international systems (Ortner 1984). By restoring human agency to a central place in social scientific investigation and by demonstrating that individuals and institutions in the Third World have a substantial impact on modifying international pressures, these efforts also helped correct the excesses of dependency and world system theory that were so in vogue during the 1960s and the 1970s (Ortner 1984). In the extreme, these paradigms proposed an all-powerful metropolitan capitalism as the explanation for underdevelopment in the periphery—in effect denying that local initiative and local response had any significant role in the making of history (Mintz 1977).

These shifts within anthropology have raised a number of theoretical concerns related to the general processes involved in sociocultural change and in the nature of integration of systems, as well as more specific methodological issues regarding appropriate levels and units of analysis. In their book, *Micro and Macro Levels of Analysis in Anthropology,* Pelto and DeWalt stress that social scientists have augmented their efforts to understand the relationships between relatively small scale and larger scale processes (i.e., between micro- and macro-level phenomena) and that this escalating concern has brought about the need to develop appropriate methodologies to guide research (1985: 187). Two important methodological responsibilities are clear delineation of conceptual models and of systems of postulated relationships among levels of analysis, and precise definitions of the relevant factors that are used as the specific linkages to articulate those levels (DeWalt and Pelto 1985: 187). Also of significant concern are appropriate sampling strategies and procedures at each level. While such considerations may be unimportant to anthropologists working solely at the community level, they become crucial when the results of research conducted at the local level are applied to larger regions or processes, because of the serious questions regarding representativeness that they raise. Also problematic, is accomplishing these methodological tasks while not diminishing the strengths of the ethnographic and other techniques already employed by anthropologists. Participant observation and key informant interviewing must be used in combination with other techniques that establish multilevel linkages. Often, this leads to more interdisciplinary approaches in which data and analysis from other social or natural science disciplines are incorporated with anthropological work. This is especially the case in research that deals with issues of contemporary policy importance.

The theoretical and methodological changes which took place throughout anthropology, also occurred within ecological anthropology. While, earlier human ecological studies by anthropologists stressed the detailed analyses of relatively small populations in local areas (e.g., Netting 1968, 1981; Rappaport 1967), recent efforts have emphasized the need to expand local level studies to the region, and even to the world, in a systemic and systematic manner (e.g., Bennett 1976, 1985, 1990: 351, Moran 1986, 1990). This is especially the case, if anthropologists are to enlarge their participation in interdisciplinary research and policy which confront global environmental problems (Bennett 1990; Moran 1990; Rappaport 1990). Important ways in which ecologically oriented anthropologists have expanded their studies beyond the single community, include the use of remote sensing, image processing, and Geographic Information Systems (GIS) which allow the integration of diverse kinds of spatially recorded data (demographic, social, economic, and ecological, as well as environmental).[2] Although these technologies have been used in other disciplines for the past two decades, the low level of formal training in spatial analytical methods and the steep learning curves of many GIS software packages, for most anthropologists, have suppressed their use within the discipline (Aldenderfer 1992: 14). However, recent interest and opportunities for training, as well as the capability of anthropologists to verify data obtained from remote sensing devices through fieldwork (i.e., through ground-level observation or "ground-truthing"), have amplified the number of anthropologists utilizing such technologies. This surge of interest led several participants in a conference at the University of California, Santa Barbara in early 1992 (*The An-*

thropology of Human Behavior Through Geographic Information and Analysis: An International Conference) to speak of an imminent "GIS revolution" in anthropology (e.g., Conant 1992: 15).

At the same time that the global awareness of the ties between development and the environment grew and consequential theoretical and methodological advancements took place within anthropology, a serious disagreement evolved concerning the ultimate destiny of Latin American peasants in the context of the spread of contemporary capitalism known as the *campesinista/decampesinista* debate. Although peasants always have been linked to larger economic systems, the expansion of capitalist agriculture throughout the Third World in recent decades has transformed the peasantry and made it reliant on wage work in labor markets tied to the world economy as never before (Smith et al. 1984). In Latin America this transformation has taken many forms, but the overwhelming trend has been towards changing subsistence farmers into wage laborers (de Janvry 1981). The overall effect has been that few rural households persist independent of wage labor, while the majority generate household income from a wide array of economic activities both on and off the farm: agricultural production, commodity production and vending, service jobs, and wage work (Warman 1981; Lehman 1982; Roseberry 1983; Collins 1988). The dependence on off-farm income is especially significant among smallholder farm families who derive the majority of household income from off-farm sources (Deere and Wasserstrom 1981). This unparalleled proletarianization raised important questions about the nature, the diversity, and the ultimate outcomes of the transformation: What forms does the process take? How stable are the emerging part-time farms, can they continue to exist and, if they do, can they do so without compelling neighbors to become rural wage earners? Such questions have provoked vigorous debate, a wide range of opposing arguments, and a number of contending positions that have theoretical significance and also affect agricultural development policy (Heynig 1982). At one extreme are the various proponents of the peasantization school (the *campesinistas*) who share the conviction that the peasant form of family production is congruous with the expansion of rural capitalism and, moreover, actually facilitates capitalist penetration (See Stavenhagen 1977; Warman 1981; Lehmann 1982). At the other limit are the supporters of the depeasantization or the proletarianization school (the *decampesinistas*) who argue that capitalist penetration into rural areas inevitably transforms peasants into landless wage workers and thus leads to social differentiation and to the emergence of a class of rural proletarians (Bartra 1974). Both these extreme positions have been criticized for being unable to capture the complexity of the process of proletarianization: the peasantization school for failing to recognize the degree to which the expansion of capitalist agriculture has been accompanied by growing social differentiation, by increasing proletarianization in the production process, and by expanding numbers of farm households who rely on semiproletarianization to survive; and the proletarianization school for being unable to explain why the emergence of a full-time proletariat appears so limited and why peasant production persists, is reproduced, and continues to be an important source of subsistence for large parts of the rural population (de Janvry and Vandeman 1987).

In their comparison of international patterns of proletarianization in agriculture, de Janvry and Vandeman observe that although there appears to be expanded proletarianization of labor in the agricultural labor process, a class of full-time rural proletarians emerges very slowly (1987). Impediments to full proletarianization range from a variety of social factors such as the flexibility of semiproletarianised farmers that allows them to participate in seasonal work and their worth to capitalist enterprises which are able to hire them cheaply because they maintain other sources of income (Maclachlan 1987), to ecological factors such as the protection of small farmers from expropriation because they occupy environments unsuitable for large-scale, heavily capitalized endeavors (Brush 1987). Following Stavenhagen (1978: 27–37), Heynig maintains that contending forces, some stimulating proletarianization and others enhancing peasantization, often act simultaneously, and suggests that embracing an uncompromising position at either extreme in the debate implies either an oversimplification of reality or a spurious dilemma (1982: 113–40).

More positively, Roseberry (1983, 1989) and others (e.g., Favre 1977; Holmes 1983; G. Smith 1989) have proposed the augmented study of the processes associated with proletarianization in order to understand the nature of agrarian transformations—placing special emphasis on the linkages among the various actors involved in the process. Of these various endeavors the reconceptualization of the peasantry as proposed by Deere and de Janvry (1979) is especially relevant. Rejecting efforts to define the peasantry as a distinct economic or sociocultural category, Deere and de Janvry undertook to develop a conceptual framework for the analysis of peasant households that considered their participation in diverse income generating activities. They argued that the analysis of contemporary peasants must be founded on the numerous relations of production in which peasant households participate (Deere and de Janvry 1979; Deere 1990). Given this conceptualization, the income generating strategies of peasant households—their survival strategies—encompass the links by which they struggle to reproduce themselves as units of production and reproduction and can be used to examine how such households persist in contexts of growing impoverishment.

Major Perspectives for Linking Development, Population, Poverty, and Environmental Destruction

The various points of view regarding the connections among development, population, poverty, and environmental decline in tropical areas of the Third World can be grouped into three general perspectives: Malthusian and neo-Malthusian; neoclassical economic and/or technological; and dependency and ecological Marxist. The issues that are raised, the specific questions that are asked, the factors that are determined to be relevant to the explanation, the relative priorities of those factors, and the proposed solutions to human and environmental problems vary with each approach.

Malthusian and Neo-Malthusian Approaches

The most familiar perspective used to explain poverty and environmental destruction in the Third World is the Malthusian or neo-Malthusian, which views mounting demographic pressure on natural resources as the paramount factor (Ehrlich 1968; Eckholm 1976; Brown 1987; Ehrlich and Ehrlich 1970, 1990). Proponents of this perspective contend that the carrying capacity of the earth is finite and that resource destruction results when too many people intensify their efforts to extract food and other needs (e.g. Ehrlich 1968; Ehrlich and Ehrlich 1990). They point to increasing populations and declining resource productivity of agriculture in many regions of the Third World and argue that this stimulates increased migration and human settlement in tropical forest regions. The results include destruction of biodiversity and natural resources. For these individuals, reducing population growth is necessary in order to arrest environmental degradation.

In opposition to the Malthusians and neo-Malthusians are the "Cornucopians" for whom population growth presents little threat and may even stimulate economic development and bring about higher standards of living through improved technology and augmented productivity (Tierney 1990: 52). These individuals, led by the ideas of Julian Simon (1980, 1981), believe that as a particular resource becomes scarce, human innovations will occur that will produce more of the resource or find an acceptable, and perhaps better, substitute (Simon and Kahn 1984). They draw on the ideas of Ester Boserup (1965) who argued that increasing population pressure gives rise to its own solution: as land becomes scarce, for example, people intensify their use of it.[3] Hayami and Ruttan (1985) and Pingali, Bigot, and Binswanger (1987) have elaborated this model and argued that a change in one factor, such as land, will lead to conservation of that factor and increased use of the more abundant factors—in this case, labor (see also Lele and Stone 1989). According to them, although increasing population may lead to difficulties in the short-term, in the long-term, a larger population will lead to innovation, technological progress, and greater productivity.

An additional criticism of the neo-Malthusians is that their arguments largely have taken place at a global or theoretical level without sufficient empirical evidence. A National Research Council study committee, for example, could find little empirical work on the linkage between population growth and environmental degradation (1986) but nevertheless concluded, "that slower population growth might assist less-developed countries in developing policies and institutions to protect the environment" (National Research Council 1992: 78). Likewise, a recent study by the United Nations Population Fund notes that although there has been considerable theoretical debate about the alleged connections among population growth, poverty, and environmental degradation, the relationships involved have not received adequate attention in the way of rigorous analysis and detailed, empirical documentation, nor have they been adequately addressed through policy initiatives (UNFPA 1991).

Studies, based on empirical research which moves beyond abstract theoretical discussions, have found several sets of linkages among population, poverty, and environmental destruction (UNFPA 1991). These analyses come from diverse perspectives (e.g., Durham 1979; Murdoch 1980; Ellen 1982; Lappe and Shurman 1988), but share the common view that environmental problems have their basis in the structure of rural poverty rather than in population increase *per se*. These studies look further than the risk of Third World populations ravaging resources to ask *why* such populations continue to grow at high rates. They identify an interacting array of economic, social, and cultural factors that maintain high levels of fertility—including the low status of women, high rates of infant and child mortality, the vital economic contributions of children, and the lack of old-age security. They contend that the decision to have children is a rational response to existing conditions and, therefore, that analysis should not center on gross correlations between population characteristics (such as growth or density) and poverty but rather on the historical and political-economic contexts that influence power and reproductive choice.

A particularly germane appraisal of the argument that overpopulation invariably explains resource scarcity is Durham's (1979) classic study of the root causes of the "Soccer War" between El Salvador and Honduras. In 1969, the government of Honduras expelled several thousand Salvadoran immigrants, many of whom had lived in Honduras for over a generation. El Salvador retaliated by invading southern Honduras. This so-called Soccer War (because it occurred shortly after Honduras lost a World Cup Qualifying Match to El Salvador in Tegucigalpa) was widely attributed to "population pressure"—the competition between Hondurans and Salvadorans for increasingly scarce arable land. Many analysts concluded that a Malthusian scenario was being played out, in which the population had exceeded the carrying capacity of the land.

Durham's study of this situation, which included empirical research in both countries, demonstrated that it was the use and distribution of land, more than population growth, density, or carrying capacity, that resulted in the problems of food production and the inability of families to meet subsistence needs. Durham found that it was the landless and landpoor agriculturalists unable to rent land in El Salvador who comprised most of the migrant stream to Honduras. Mostly renters and sharecroppers, the Salvadorans' access to land depended on the decisions of large landholders rather than on competition with Honduran smallholders. In fact, immigrants and poor Honduran farmers joined forces to challenge a large hacienda owner who attempted to incorporate national lands into his estate. Durham concluded that the land base of poor farmers decreased to the point of threatening survival only partly as a result of population increase. As he put it, "Land use patterns show that land is not scarce for large landholders" (1979: 54).

Neoclassical Economic Perspectives

The many variants of this view share the assumptions that well-functioning markets are always the best means of allocating natural resources and that competition necessarily leads to appropriate management. Consequently, under a system of private property, individual maximizers will rationally manage resources to their best advantage in order to remain competitive within the market. In this framework, environmental degradation in the tropics is often attributed to faulty incentive systems affecting economic and demographic behavior centered around the use of common property resources (Hardin, 1968, 1977; Clark, 1974) and to the "irrational," "traditional" (i.e., stagnant and using "primitive technology") land-use decisions of small producers (National Resource Council 1982). From this perspective, solutions to environmental degradation are based on modifying the internal organization of production through the privatization of landholdings (Hardin 1968) or through "appropriate" interventions such as the introduction of "modern" technologies (Hardin and Baden 1977).

Critiques of this perspective have included arguments aimed at debunking the myths and misconceptions about communal resource management (Dove 1983; McCay and Acheson 1987; Feeney et al. 1990); at more objectively evaluating the human and ecological sustainability of smallholder agricultural, agroforestry, and pastoral systems (Nations and Nigh 1980; Dove 1983; Spooner 1987); at demonstrating the rationality and adaptive behavior of small farmers and peasants (Bennett 1969; Brush 1977, 1987; DeWalt 1979; Barlett 1980); and at pointing out the relatively greater importance of external factors, such as government policies, class position, and land tenure, to the internal organization of resource management (Murdoch 1980; Hecht 1981; Grossman 1984; Collins 1986; Little and Horowitz 1987; Painter 1987; Stonich 1989; 1991c).

Dependency and Ecological Marxist Approaches

In contrast to neoclassical economic perspectives that emphasize factors internal to national economies, the wide variety of dependency perspectives focus on the external factors that alter production systems that, in turn, induce environmental decline (Murdoch 1980; Smith and O'Keefe 1980; Eckholm 1982). Some earlier variants of this point of view emphasized export production for the global market and the imposition of inappropriate management practices and technologies by international corporations as the most important causal factors. Environmental destruction in the Third World is viewed as the result of the imprudent use of management practices and technologies developed for different environments (e.g., Janzen 1973) or of the excessive demand in the developed world for environmentally destructive commodities (e.g., Myers 1981; Nations and Komer 1983). Criticisms of the application of such dependency approaches to environmental issues are similar to more general evaluations of dependency and world-system models: i.e., they are overly simplistic and cannot adequately explain continuing underdevelopment or variations in development performance (Cardoso 1977; C. Smith 1980).

A more recent and more thorough perspective has been advanced by individuals such as Ernesto Leff (1986) and James O'Connor (1988) who are attempting to expand traditional neo-orthodox Marxist analysis to a broader "Ecological Marxism." Their analysis of the Latin American crisis focuses on the contradictions between the forces and relations of production on one hand and the ecological conditions of production on the other. They examine, both theoretically and empirically, the tendency of capitalist development to destroy its own ecological conditions of production—the environmental basis for maintaining dependent capitalism—thereby aggravating economic and social crisis in the long-term (e.g., Leff 1986; O'Connor 1988, 1989; Karliner 1989; Faber 1992a, 1992b).

Political Ecology:
Political Economy + Human Ecology

The argument in this volume is that although several of the preceding approaches (especially that of the ecological Marxists) identify one or more factors or linkages that are germane to a comprehensive explanation of poverty and environmental deterioration in areas of the Third World such as southern Honduras, none are adequate by themselves. As an alternative, the overall approach here is a more inclusive framework that integrates political-economic and human-ecological analysis. The political-economic analysis examines the interacting roles that social institutions (international, national, regional, and local) play in providing constraints and possibilities that affect human decisions that in turn affect those institutions as well as the natural environment. Human ecological analysis allows the consideration of crucial demographic factors, environmental concerns, and issues related to human health and nutrition. It expands the perspective of political economy to include a systematic examination of the distribution and use of resources and the dynamic contradictions between society and natural resources.

The integration of ecology (or at least environmental issues) with political economy has emerged as one of the major frameworks used to understand environmental destruction in various parts of the Third World (e.g., Blaikie 1985, 1988; Blaikie and Brookfield 1987; Little and Horowitz 1987; Redclift 1984, 1987). This integrated perspective has been termed "political ecology" and has been used in a variety of disciplines to demonstrate how interconnected social, economic, and political processes affect the way natural resources are exploited (See e.g., Messer 1987; Schmink and Wood 1987; Bassett 1988; Sheridan 1988; M. Chapman 1989 Johnson 1989; Saldanha 1990). The use of the expression, "political ecology" has some history in anthropology: Eric Wolf used the phrase in his response to the symposium *Dynamics of Ownership in the Circum-Alpine Area*, published in 1972. As a discussant for that symposium he called for augmented research that "... combine our inquiries into multiple local ecological contexts with a greater knowledge of social and political history ..." and "... the study of inter-group relations in wider structural fields ..." (Wolf 1972: 204–5). Although using the term, Wolf did not elaborate on political ecology as a concept, a theoretical perspective, or as a methodological approach.

A more recent and the most elaborated explanation of political ecology, from the point of view of the discipline of geography, was advanced in 1987 by Blaikie and Brookfield in their book *Land Degradation and Society* and further detailed in an article by Blaikie in 1988. Political ecology as a theoretical approach laid out by Blaikie and Brookfield (1987) includes the following essential elements: (1) Political ecology combines the concerns of ecology and political economy thus integrating human and physical approaches to environmental destruction (p. 17); (2) Analysis follows a "... chain of explanation ..." through different scales (levels of analysis) beginning with the decisions of local land managers (such as farmers), the interrelations among local managers and other groups in society who affect local land management, and as well as the roles of the state and world economy (p. 27); and (3) Because political economy insinuates analysis of structures external to local groups which affect options and decisions, considerable attention is focused on the ways in which international capitalism and the state affect natural resources and local people (Blaikie 1988: 141).

As presented by Blaikie and Brookfield, political ecology embodies a broad approach that integrates a hierarchy of scales and a variety of methodologies. Their concern is with environmental degradation in its broadest meaning. Their goal is to establish a theoretical approach that explains *why* the decisions of land managers sometimes cause environmental deterioration which subverts their own livelihoods. As such the approach is potentially useful in explaining the persistence of seemingly irrational land management decisions which may lead to temporary declines in productivity as well as permanent environmental degradation. Also of use is Blaikie's (1985: 125) discussion of the concept of eco-demographic or spatial marginality in which small producers are displaced to less fertile or environmentally more vulnerable locations as a result of land expropriations by the state or by large agribusiness concerns. Small producers are placed in a position where they may be forced to over-exploit a scarce resource in order to survive.

While the political ecology approach as laid out by Blaikie and Brookfield represents an advance over other less inclusive explanations of environmental degradation, the framework is not complete. Although Blaikie and Brookfield emphasize the relationship between society and resources and among various levels or scales, the political economy is described as "exogenous" when the process of land-use decision making is represented (Blaikie and Brookfield 1987: 70). Moreover, in common with earlier "world system" and "dependency" perspectives, there seems to be little attention given to the ways in which local people mitigate the impact of external forces and how such local responses, in turn, affect the wider political economy. In addition, demographic dynamics are included only obliquely in the analytical framework.

The research approach used in this book augments the framework provided by Blaikie and Brookfield by systematically examining the ways in which local actors mediate the impact of external forces and by integrating crucial demographic factors (especially population growth, density, and distribution) into the analysis. Greater attention is given to the differential consequences of exterior pressures on varying classes and interest groups and to the ways in which human agency and local level initiatives influence the broader political economy.

This study owes a great deal, as well, to the "human systems ecology" paradigm developed by John Bennett (1976, 1980, 1985, 1986, 1990). The term, "human systems" ecology, was introduced to distinguish John Bennett's ecological paradigm from other forms of cultural and human ecology. The use of the word "system" between "human" and "ecology" emphasizes Bennett's interest in the institutional arrangements which mediate between human beings and the natural environment. Bennett's paradigm, which views human and physical factors as a single system, includes a focus on regional systems in which diverse human groups adapt in predictable ways to environmental resources; to one another; to hierarchical market and administrative forces; and to pressure groups and other forms of quasi-organized social and political interests (Smith and Reeves 1989). A major achievement of Bennett's ecological perspective is that it brings to bear the concern for studying complex socioeconomic processes with a flexibility of analysis which takes into account micro-level decision making as well as institutional pressures and facilities—all of which must work with the available material resources. It allows the clarification of the linkages between the local specifics of human thought and behavior and macro-level institutions such as markets and government agencies. Bennett's approach is particularly useful in comprehending the complex, multi-level workings of proletarianization through delimiting the multiple factors encouraging and limiting the process and then attempting to determine the systemic interrelationships among them.

Analytical Framework

The integrated perspective used in this study demonstrates the close linkages between social processes and environmental deterioration in southern Honduras. The goal of the methodology was to provide a systemic framework in which to articulate the relevant factors (demographic, environmental/ecological, social, and economic) and levels of analysis (individual, household, community, regional, national, and international) affected by the imposition of the dominant development model (see the Appendix, especially Figure A.1, which specifies the relevant factors and levels of analysis as well as the sources and kinds of data associated with each). The ultimate objective was to integrate a representative sample of micro-level studies of individuals, households, and communities into a hierarchical framework composed of multiple and increasingly macro-levels of analysis (the region, the nation, and the world). A further aim was to add a historic dimension by positioning the traditional foci of anthropology—individuals and communities—at the convergence of local and world history. The underlying assumption, that variability existed in terms of all factors and at all levels, required that sampling strategies be chosen so as to be capable of collecting data that were representative of that heterogeneity.

The ensuing chapters use a political ecology approach to examine the links among development, population, poverty, and environmental destruction in southern Honduras. In general, the study moves from more macro to more micro levels of analysis, linking levels as the presentation proceeds while also discussing germane demographic, social, economic, and ecological factors at each stage. Analysis begins by placing southern Honduras within the national context and by examining the interconnections among regional environmental and demographic factors (Chapter 2). Still centered on the region, the historic dimension then is integrated into the analysis with special emphasis on the key processes of change and their significance (Chapter 3). Next, patterns of regional development and associated changes in the political economy are explicitly linked to ecological changes and environmental destruction (Chapter 4). The focus of analysis moves to the municipality and the community in Chapter 5, in which significant demographic, social, economic, and ecological factors are linked to the regional processes previously presented. Continuing the process of proceeding from macro- to more micro-levels of analysis, the next chapter (Chapter 6) focuses on the household and examines variations in household economic survival strategies—including both on-farm and off-farm activities—and presents an overall model which links the various factors and processes examined throughout the work. Finally, the political ecology model and levels of analysis are linked to broader processes of social and environmental change within Central America and to their policy implications.

Notes

1. The increasing involvement of anthropologists in addressing global environmental problems is suggested, in part, by the following: the establishment of the Environmental Task Force by the American Anthropological Association and the Commission on Human Rights and the Environment by the Society for Applied Anthropology; the creation of two centers for training, The Human Ecology Laboratory at Hunter College and the Anthropological Center for Training and Research on Global Environmental Change at Indiana University; the involvement of anthropologists at the three National Centers for Geographic Information and Analysis (NCGIA) at the University of California-Santa Barbara, the State University of New York-Buffalo, and the University of Maine; and the preparation of several student training manuals.

2. For an introduction to the use of remote sensing and GIS by anthropologists see Conant 1990, 1992: 15; Aldenderfer 1992: 14; Goodchild 1992: 14–15; and Moran 1992: 16. For collections of conference papers see Behrens and Sever 1991; Aldenderfer and Maschner n.d.

3. Boserup (1965, 1981) suggested that the biophysical system is far less dependent on resources and on the state of technology than the Malthusians and neo-Malthusians would suppose. Boserup demonstrated that output from a given area can increase far more from the adoption of more intensive production systems (i.e., through increased inputs from labor or innovations in technology) than neo-Malthusian models would assume. In spite of this divergence, Boserup, like the neo-Malthusians, segregates population as a causal variable and, moreover, views population growth as a primary basis for technological change in agriculture (Boserup, 1965: 56). Redclift (1987: 31) argues, moreover, that the evidence of diminishing returns presented in her own books suggests that Boserup's contribution has been to rectify rather than to disprove Malthus.

2

Southern Honduras:
Environment and Demography

There are two seasons here (southern Honduras) hot and hotter. ... During one it's hot and dry and during the other hot and wet.

—Southern Honduran peasant, 1982

You know what they say about the south? ... Going to the south is good practice for going to hell.

—Honduran government official, 1981

Southern Honduras is a triangular-shaped region covering approximately 5,775 square kilometers in the southern-most part of the country along the Gulf of Fonseca (Figure 2.1). It is located primarily in tropical dry and subtropical moist forest zones and encompasses a coastal ribbon of mangrove and wetlands, a lowland plain, and rugged foothills and mountains (Holdridge 1962). Steep slopes, irregular precipitation, and easily eroded soils make the area extremely susceptible to environmental destruction as well as highly risky for agriculture (USAID 1982; SECPLAN/USAID 1989).

Most often the south is delineated as a "region" on the basis of a combination of environmental and political boundaries: on the north by a portion of the Central American Antilles chain of mountains, on the south by the Gulf of Fonseca, on the west by the Rio Goascoran, which forms the national border with El Salvador, and on the east by the Rio Negro, which constitutes part of the frontier with Nicaragua. All demarcations of the region include the entire political departments of Choluteca and Valle, which together account for at least 90 percent of the surface area and the population in all regional definitions.[1] The two departments of Choluteca and Valle are divided geopolitically into twenty-five municipalities—sixteen in Choluteca and nine in Valle—each of which is subdivided into smaller towns and hamlets (Figure 2.2). Municipalities can also be divided on the basis of their prevailing elevations into the nine municipalities that lie primarily in the lowlands and the sixteen that are found in the highlands.[2] This work is based on research carried out in the north-central portion of the region, including the contiguous municipalities of Pespire, Nacaome, San Lorenzo, and the northern part of Choluteca. The communities which are the basis for the in-depth local, house-

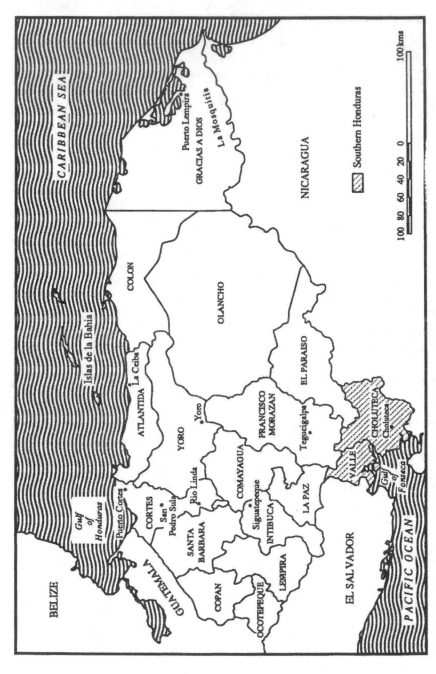

FIGURE 2.1 Honduras and Southern Honduras. Source: USAID 1982.

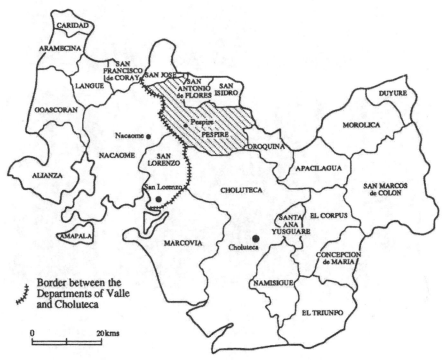

FIGURE 2.2 Southern Honduras and the Municipality of Pespire. Source: CSPE/OEA 1982

hold, and individual level analyses found in Chapters 5 and 6 are located in the highland municipality of Pespire. All three of the region's largest urban centers (Choluteca, San Lorenzo, and Nacaome) are found in the lowlands, part of the Pacific coastal plain.

Southern Honduras is a land of potential and paradox. Until a few generations ago, the region was rich in natural resources, including abundant pine, oak, and mangrove forests and a wide coastal plain with fertile soils. In recent decades, however, human and environmental conditions deteriorated significantly, and most of the population now face severe resource constraints and live in chronic poverty. Simultaneously, pollution and environmental degradation devastate the natural resources of the region, decreasing their productive potential for current and future generations, negatively affect human health, and threaten the existence of myriad plant and animal species. The trees that once covered the mountainous areas are nearly gone, cleared by small farmers desperate for land on which to plant food crops, by larger farmers anxious to create pasture for cattle, and by the southern people in general, who rely almost exclusively on fuelwood for their energy needs. The bare slopes show the ravages of erosion which has precipitated the widespread degradation of the region's watersheds. Extensive watershed deterioration,

"Rich" highland peasant directing a group of twelve hired agricultural laborers who are intercropping corn and sorghum in his field. Laborers are using
digging sticks; corn and sorghum seeds are held in containers at their waists.

emanating in highland areas, has escalated the siltation of rivers and of mangrove forests along the coast. Increased sediment loads, in turn, have combined with agricultural runoff contaminated with pesticide residues from
large-scale, export-oriented farms in the lowlands. In addition, the excessive
harvesting of mangroves for use as domestic fuelwood and in the salt-
extraction industry and the significant expansion of shrimp farms have accelerated the destruction of coastal environments as well (Stonich 1992).

Prior to World War II, the region was distinguished by moderate stability
between medium-sized cattle ranchers in the lowlands and small-scale subsistence farmers in the highlands (Boyer 1982). After the war, the region entered a
period of rapid economic development, driven in large part by U.S. development assistance and the Honduran state, and became incorporated into the expanding worldwide market for agricultural commodities, (Stonich and
DeWalt 1989). Propelled by the Alliance for Progress in the 1960s, large-scale
cotton and cattle farming grew to dominate the area, and the richest farmland
became dedicated to large cotton estates and pasture. The widespread social
and ecological problems that accompanied the cattle boom are well documented (DeWalt 1985; Boyer 1987; Stonich 1989). Although the majority of the
expansion took place in the lowlands, smaller cattlemen in the highlands expanded their pasture land, often at the expense of kin and neighbors (Boyer
1987: 14–15).

The impact of cotton was confined to lowland areas of the departments of Valle, Choluteca, and neighboring Olancho, but in these zones its effects were no less significant. To create the large cotton estates, many small-scale farmers were driven from their land both by legal and illegal means (Parsons 1976; White 1977). With cotton came not only social dislocation and the increasingly inequitable distribution of land and wealth, but also serious ecological disturbance, in part due to the excessive dependence on chemical inputs (Murray 1991). Since its inception, the cotton industry went through several cycles of expansion and contraction, but began a decline after 1980 from which it has not recovered.

In the 1980s, as part of the promotion of nontraditional exports throughout Central America, national and foreign investors began financing two new nontraditional export commodities in the south—melons and cultivated shrimp—which grew 23 percent and 22 percent, respectively, between the mid-1970s and the mid-1980s (USAID 1990: 1–2).[3] In 1989–1990 these two commodities contributed an estimated US$25 million in export earnings to the Honduran economy (Meckenstock et al. 1991: 4), but as with the earlier booms in cotton and cattle, these earnings have been offset by both environmental and social costs.

Increasing the pressure on the region's natural resources is a population with the highest total fertility rate (8.95) in the nation (Howard-Borjas 1990: 9), with more than 50 percent of inhabitants less than 15 years of age (estimated from community level studies presented later in this volume) and with an annual rate of growth that remains at approximately 2.1 percent per year despite an out-migration rate of 1.3 percent annually (Stonich 1989). The south remains the most densely settled region of the country, comprising only 5.8 percent of the total national area but containing approximately 9.3 percent of the population with a population density almost twice that of the country as a whole (Stonich 1989). By 1974, the region was characterized by the highly unequal distribution of landholdings due to the ruthless expansion of export commodities and the growing human population, in which 68 percent of producers had access to less than 5 hectares of land—a total of only 10 percent of the total land area—while 1.6 percent of landowners controlled 47 percent of the land area (Stonich 1986: 143).

Worsening rural destitution accelerated migration to the region's largest urban center, Choluteca, whose population, of approximately 55,000 in 1988, more than doubled since 1974 (SECPLAN 1988, 1989). People from degraded areas in the south also make up a significant proportion of the migrant stream to the capital city of Tegucigalpa, the industrial centers of San Pedro Sula and the North Coast, and the relatively unpopulated areas of *La Mosquitia*, Honduras' last remaining significant area of tropical humid forest (Stonich 1989).

It is not difficult to understand why southern families have chosen to "look for a better life" elsewhere than in the environmentally degraded areas of the rural south. In terms of most economic, nutritional, and health measures, the people of southern Honduras fall below the national averages. Southern Honduras is more dependent on agriculture than is the rest of the country, with approximately 70 percent of the population directly relying on agriculture for

their livelihood; yet two-thirds of all farms in the region are less than 5 hectares in size, too small to provide sufficient income for the average rural family
(Stonich 1991a). Moreover, approximately 40 percent of the region's families
are landless and must rely solely on income earned through wage work
(Stonich 1991a). The combined effects of dwindling landholdings, insufficient
off-farm employment opportunities, and the unequal distribution of resources
between rural and urban populations and within the rural sector resulted in
an estimated 70 percent of rural families living on less than US$20 per month
by 1980 (CSPE/OEA 1982: 112–114).

Results from the study of nine highland and lowland communities in 1982
showed that 65 percent of the children under 60 months of age were stunted
(i.e. below 95 percent of the standard height-for-age ratio recommended by
the World Health Organization [WHO]), and 14 percent were wasted (i.e. below 90 percent of the standard), and more than half of all families failed to
meet energy (calorie) requirements in some communities (DeWalt and Dewalt
1987: 39). A more recent 1987 study of children under 60 months of age estimated that 50.6 percent of children in the south and 44 percent nationally were
stunted (Howard-Borjas 1990: 10). A regional analysis of the results of the National Nutrition Survey conducted in 1986 indicated that 35 percent of all first
graders in the south were more than two standard deviations below the
height-for-age ratio of the WHO reference population—indicating widespread moderate to severe degrees of chronic undernutrition for slightly older
children as well (Stonich 1991b: 58). In the early 1980s, when the infant mortality rate was estimated to be 78.6/1000 in Honduras and 89/1000 in the south
(Howard-Borjas 1990: 9–10), community level studies in rural areas suggested
even higher rates—averaging 99/1000 and reaching 117/1000 (Stonich 1986:
196; DeWalt and DeWalt 1987: 39). Both undernutrition and infant mortality
were related to the inability of farm families to gain access to enough land or
non-farm employment to sustain themselves (Durham 1979; DeWalt and
DeWalt 1982; Stonich 1986, 1991b).

In addition to low income and widespread undernutrition, the people of
southern Honduras are faced with significant risks to health. In interpreting
these data, it is important to remember that Honduras ranks at or near the bottom in terms of most indicators of health, compared to the other countries of
Central and Latin America (PAHO 1990). According to a report by the Honduran Secretary of Public Health in 1988, residents of the southern region exhibited rates of respiratory infections and diarrheal diseases above the national
averages, while the incidences of malaria and dengue fever were the highest
in the nation (SECPLAN/USAID 1989: 270). In addition, land and water contamination from the excessive reliance on pesticides in export agriculture, as
well as high levels of pesticide residues in food supplies, have generated serious public health problems. Water quality data from cotton growing areas in
the lowlands showed heavy contamination from DDT, dieldrin, toxaphene,
endrin, and parathion by the mid-1970s (ICAITI 1977). A 1981 study to determine the levels of pesticide poisoning in the area around the city of Choluteca
revealed that approximately 10 percent of the inhabitants had pesticide levels
sufficiently high to be considered cases of intoxication (Leonard 1987: 149). A

subsequent 1986 study of pesticide contamination of ground and surface water in the department of Choluteca found high levels of DDT and other organochlorine pesticides commonly used in cotton from the 1950s into the 1980s (Buseo et al. 1987), and pesticide poisonings from both insecticides and herbicides were the second leading cause of death in cattle grazing in cotton growing areas (Aguilar 1988). Although most are no longer used in the south, their effects on the physical environment and on human health are just beginning to be recognized (Murray 1991). The problems with growing pesticide use in producing nontraditional crops, especially melons, is also evident. Although inadequately reported, pesticide related illness rates have been high in Choluteca due to heavy pesticide use in cotton since the 1950s (Murray 1991). But as cotton production declined in the 1980s, the incidence of pesticide related illnesses did not diminish, and, in fact, hospital entry reports increased by 14 percent between 1981 and 1985 (Buseo et al. 1987). The array of environmental, natural resource, and human problems in the south presented above do not occur in isolation from each other but are closely interrelated, as shown more fully below and in subsequent chapters.

The Environment of Southern Honduras

On the Pacific coast, Honduras shares the Gulf of Fonseca with El Salvador and Nicaragua. The portion of southern Honduras lying on the Pacific Ocean is approximately 110 kilometers wide at its narrowest point, but because the coast winds around the Gulf of Fonseca, there are approximately 145 kilometers of coastline. The coastal zone extends over approximately 1,000 square kilometers (16 percent of the surface area of the region) and is composed of mangrove forests, wetlands, and the volcanic cone islands in the Gulf. The area is fed by five major river systems (the Goascoran, Nacaome, Choluteca, Sampile, and Negro rivers) that drain approximately 13 percent of the country. The youngest geological subregion, it is composed of recent sediment and, as a transition zone between land and sea, it is an ecologically vital area (SECPLAN/USAID 1989).

The southern lowland plains extend fan-like beyond the coastal mangrove ecosystems. They are one of the few extensive plains on the Pacific coast of Central America—averaging 40 kilometers in width, 162 kilometers in length, and covering approximately 1,450 square kilometers (22 percent of the surface area of the region). The lowlands are not homogeneous but are composed of two terraces. The lower terrace extends over a 780 square kilometers area from the upper limits of the mangrove forests to a height of approximately 10 to 15 meters above sea level. Following the course of the major rivers, it is of recent alluvial formation. The soil is generally a combination of clay and loose sand with variable drainage depending on altitude. The second terrace, covering 650 square kilometers, extends from the first terrace to the foothills. While the first terrace is a flat plain, the second is undulating and appears to have originated by the uprising of the sea floor.

The lowlands give way to steep foothills which quickly become the jagged mountain ranges that form a broad base to the northeast and constitute 62 percent of the region. The "highlands" are designated to include those foothills and mountains located over 200 meters in elevation and having a slope greater than 15 percent. Although the highest peaks occasionally reach altitudes of 2,000 meters, they generally average between 600 and 800 meters within most of the region. They are extremely rugged with slopes varying from 15 percent to 60 percent and form diverse valleys and heterogeneous micro-climatic and ecological zones.

Climate

Two wind systems interacting with the regional topography are important in regulating climate—especially in determining the distinct rainy and dry seasons which govern agriculture (Figure 2.3). For most of the year the stronger northeasterly trade winds predominate. As air moves across northern Honduras and then across the mountain range to the south, it cools and releases moisture. While this provides considerable rainfall on the windward mountain slopes, the leeward slopes and the bulk of the plateau land in the south are left in a rain shadow. From November until April, these northeasterly winds bring the dry season: there is little or no rainfall and accompanying low levels of humidity. During this period, sea breezes from the Gulf remain relatively weak; however, sometime between the end of April and the beginning of May, when the land temperature is at its highest and the difference between water and land temperature is greatest, these southerly winds become stronger and penetrate farther north into the southern region. There they collide with the northeast winds and form frequent and sometimes violent thunderstorms—the rainy season. Southerly winds predominate until approximately the middle of July, when once again the northeast winds prevail and a brief hiatus (the *canícula*) in the rainy season occurs. Whether or not the *canícula* takes place in any given year, as well as its length, is highly variable. It may not happen at all, or it may last for a month, after which time southerly winds again prevail until October, bringing with them a return of the rainy season.

This combination of factors results in considerable annual, seasonal, monthly, and intra-regional variability in precipitation. Based on rainfall data collected from a 22 year period (1952–1974), average annual precipitation ranged from 500 millimeters in the northeast to more than 2,400 millimeters in the southwest, and drought conditions were quite common (Hargreaves 1980). Intermittent and worsening drought since the 1960s remains a constant concern for peasant cultivators who expect to lose every third corn harvest because of inadequate or irregular precipitation (Boyer 1982: 54–56).

As with precipitation, average annual temperatures decrease with altitude; highland areas have mean temperatures of approximately 21 degrees Centigrade, lowland areas approximately 28 degrees Centigrade. The south can be quite hot with maximum temperatures frequently reaching 40 degrees Centigrade during the hottest months of March and April (CSPE/OEA 1982).

Major Habitats and Environmental Problems

The southern coast is characterized by tropical marine ecosystems dominated by large expanses of mangroves and wetlands.[4] These mangroves and associ-

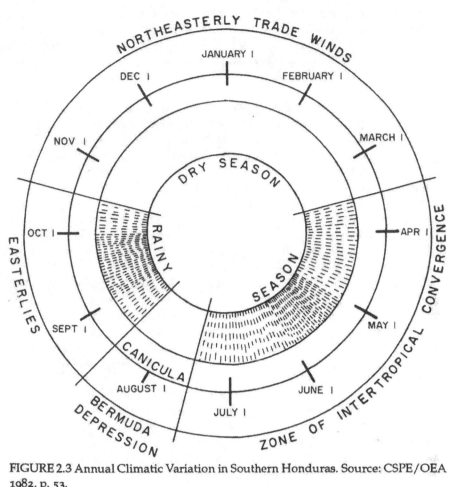

FIGURE 2.3 Annual Climatic Variation in Southern Honduras. Source: CSPE/OEA 1982, p. 53.

ated estuarine waters perform crucial ecological functions: they have high primary productivity; they serve as breeding and nursery areas for important species of shellfish and finfish; the roots of the mangroves act as dams trapping the alluvium and nutrients washed from higher elevations—the first step in new soil formation; and they stabilize the coastline protecting it from inundation. Because the geophysiography of the Gulf of Fonseca has excluded Honduras from offshore fishing grounds, there are no major industrial fisheries on the Pacific coast. The Honduran estuaries, however, provide critical habitats for the large offshore shrimp populations commercially exploited by neighboring Nicaragua and El Salvador.[5] In 1980 the Food and Agricultural Organization of the United Nations (FAO) estimated that the southern Honduran artisanal sector consisted of approximately 1,000 fishers who operated small vessels, mostly dugout canoes, of which only 25 percent were outfitted with motors. Typical gear utilized are cast nets, fish hooks, and gill/trammel nets. The major species exploited by artisanal fishers include a diverse number of finfish, two species of arcshells (*Anadara tuberculosa* and *Anadara similis*), and the mangrove oyster (*Crassostrea mangle*). Although shrimp and lobster do not reach commercial size in Honduran waters, harvesting of postlarval and juvenile shrimp, which serve as seed stock for the rapidly expanding shrimp farm industry in the south, has resulted in significant declines in shrimp populations. Indeed, as discussed in Chapter 4, the expansion of shrimp mariculture, as part of the latest trend in the promotion of nontraditional exports, constitutes the most immediate and significant threat to coastal ecosystems.

Remnants of tropical dry forest occur inland from the coastal zone. Such tropical deciduous forests are found in areas where marked seasonality of precipitation predominates and were once prevalent along the entire Pacific coastal plain of Central America. Although deciduous forest once represented the dominant vegetation type in the lowlands of the Pacific coastal region of southern Honduras as well, it has been almost completely eliminated by agriculture (crops and cattle). Only a few small remnants remain, mostly as scattered gallery forests along streams and rivers.

Pine and oak associations, corresponding to Holdridge's (1962) subtropical moist forest, occur at altitudes from 600 to 1,800 meters. Predominant species are oak (*Quercus*) and pine (*Pinus oocarpa*) at lower elevations, and pine (*Pinus psuedostrobus*) at higher elevations of the zone. Understory varies from grassy cover to low shrubs and tall grasses. Slash and burn agriculture, cattle grazing, cutting of trees for fuelwood and construction, and commercial logging of pine for export have greatly modified this habitat.

Islands of cloud (montane rain) forest are found at elevations from 1,350 to 2,300 meters where the almost daily cloud buildup and the lower evaporation rates on mountain peaks provide moisture for the lush plant growth. These highland broadleaf forests generally are surrounded at lower elevations by pine and oak forest. Cloud forests are important in the regulation of surface and ground water supplies for drinking, irrigation, and hydroelectric power production. For example, the capital city of Tegucigalpa gets a large percentage of its drinking water from cloud forests. The role of intact forests in maintaining water supplies is especially evident in these forests because condensation on leaves and trunks is the source of much of the precipitation. Because of their rugged terrain many of these cloud forests remained fairly intact until recently. However, they are being seriously degraded as increasing populations of desperately poor farmers expand slash and burn cultivation into these formerly remote areas.

Land Use and Vegetation

Vegetation in southern Honduras is related to a number of climatic, physiographic, human, and other factors; and today the region is as much a cultural landscape as a natural one. Clearing natural ground cover and planting foreign (i.e., extra-local) cultigens have occurred since prehistoric times (Stone 1957: 73) and cattle raising began shortly after the Spanish conquest (Johannessen 1963: 2)—processes affecting more recent patterns of human settlement, land use, and vegetation. Differential patterns of contemporary land use in the region are the subject of subsequent chapters but for the present can be broadly characterized: in the post–World War II period semi-subsistence agriculture generally has been located in the highlands and foothills while the cultivation of agricultural commodities for export increasingly has predominated in the lowlands. Most small producers are concentrated on the steep mountain slopes that are marginal for agriculture. Although large landholdings are relatively rare in rural highland communities (Durham 1979; Boyer 1982), there is considerable inequality in access to land among people farming the steep slopes. In these communities, agriculture is based on various systems of shifting cultivation involving interplanting corn, sorghum, and (to a much lesser extent) beans. In the past, two or three years of cultivation were followed by a period of several years of fallow, during which the soil was allowed to recover its fertility. More recently, fallow cycles have been significantly shortened or eliminated in most parts of the region. As will be discussed in later chapters, cattle raising which was first concentrated in the lowlands expanded to highland areas during the internationally and nationally promoted cattle boom of the 1960s and 1970s during which time the percentage of pasture land in the region grew to 61 percent of the land area (Stonich 1989).

According to the 1989 Environmental Profile of Honduras, the coastal zone comprises a total of 71,409 hectares: 46,758 hectares (66 percent) were mangrove forest; 13,757 hectares (19.3 percent) were salt and mudflats; 8,804 hectares (12.3 percent) were in shrimp farms; 1,297 hectares (1.8 percent) were in salt making ponds; and 624 hectares (.9 percent) were used as fishing grounds by artisanal fishers (SECPLAN/USAID 1989). Approximately 42,000 hectares of the bordering lowland areas were cultivated in sugar cane (10,000 hectares), cotton (3,400 hectares), melons (4,300 hectares), cashews (2,700 hectares), sesame (2,000 hectares), sorghum, soya, or were used to graze cattle on unimproved pasture land (SECPLAN/USAID 1989). While most of the lowlands is taken up in commercial agriculture and in pasture for cattle, the plain is dotted with *jicaro* (*Crescentia alata*), *tiguilote* (*Cordia dentada*), *ceiba* (*Ceiba petandra*), and *guanacaste* (*Entorolobium cyclocarpus*) trees. More than 40 species of broad leaf deciduous trees grow in the foothills along river courses or ravines and between fields used for subsistence production or for pasture. These fields extend up to roughly 800 meters where pine forests begin to dominate. In 1989, the principal crops cultivated in the foothill and highland areas included: maize (18,533 hectares), sorghum (18,106 hectares), and beans (417 hectares) (SECPLAN 1990).

Soils and Agricultural Potential

Data on the physical and chemical properties of southern soils are meager and inconsistent. Information about the kinds and distribution of soils are found mainly in generalized studies dealing with the country as a whole (e.g., OAS 1962) or in a few large scale soil maps of small areas. Although several systems of classifying soils have been used, they have tended to be biased toward the identification and evaluation of soils useful for capital intensive agriculture. In general, these studies have concluded that the most fertile soils occur in the lowland plains and that little of the south is "suitable" for intensive cropping or for pasture (Schreiner and Badger 1983: 14; CRIES 1984; Stonich 1986: 75–81).[6] In the classification scheme determined by Schreiner and Badger (1983), for example, there is a definite relationship between relative fertility and highland and lowland areas. As defined by CONSUPLANE (1982), the municipalities identified as having relatively fertile soils are located predominately in the lowlands while the remaining municipalities determined to have infertile soils are virtually all in the highlands (Schreiner and Badger 1983: 14).

The human and ecological consequences of this differential locus in agricultural systems and agricultural potential is a major concern in the remaining chapters. Here it is important to point out that highland areas assessed to have the lowest agricultural potential also have the highest population densities in the region. In the department of Choluteca, the two municipalities with the highest population densities in 1988 (Concepción de Mariá and Oroquina) are both located in the highlands; more than 90 percent of the soil groups found there, are rated as having a uniformly low productivity potential for all crops (CRIES 1984). Regardless of this assessment, people in these municipalities engage in the intensive production of maize, sorghum, and beans—the most important human food crops—a traditionally vital economic activity made more important because of the deteriorating economic situation in the region.

The Demographic Dimension

The rate of population growth in Honduras has been among the highest in the world, averaging 3.1 percent per year between 1950 and 1974 and rising to approximately 3.4 percent from 1974 to 1988 (SECPLAN 1988; Stonich 1989). As shown in Table 2.1, the population of the south grew notably as well, and by 1988 reached 415,449. However, after 1950, population growth in the south did not keep pace with the rest of the country due in part to the south's high infant mortality rate and to extensive out-migration from the region. Population densities in the south, however, remained significantly greater than population densities in the nation as a whole due to a fertility rate higher than the national average and to in-migration by Salvadorans prior to the 1969, Soccer War between El Salvador and Honduras.[7]

Although population increased absolutely throughout the region between 1950 and the 1980s, the spatial distribution within the region was by no means uniform, nor was the rate of growth, between highland and lowland areas (Figure 2.4). In 1950, rural population densities in the highlands surpassed those in the lowlands—approximately 36.6 persons per square kilometer in highland areas compared to 23.1 persons per square kilometer in the lowlands (Boyer 1982: 96–99). By 1974, due in large part to escalating migration from the highlands to the lowland south and to urban and rural destinations outside the region, rural lowland densities slightly exceeded those in the slower growing highlands—50.4 persons per square kilometer to 48.3 (Boyer 1982: 96–99).

TABLE 2.1 Population Growth and Population Density, Honduras and Southern Honduras: 1901 to 1988

	Honduras			Southern Honduras		
Year	Population	Index[a]	Density[b]	Population	Index[a]	Density[b]
1901	543,741	40	4.85	78,790	46	13.6
1930	854,104	62	7.62	109,341	63	18.9
1950	1,368,605	100	12.20	172,620	100	29.8
1961	1,884,765	138	16.80	230,082	133	39.8
1974	2,656,948	192	23.70	285,237	165	49.4
1988	4,443,721	324	39.05	415,449	240	72.0

[a] Index 1950 = 100.
[b] Density in inhabitants per square kilometer.

Source: DGECH 1981. SECPLAN 1989.

During the ensuing years, rural highland and lowland densities have remained fairly in balance with highland densities approximately 70 persons per square kilometer and lowland densities 67.3 persons per square kilometer in 1988 (Calculated from data in SECPLAN 1988 and 1989). In absolute terms, the lowlands contain 60 percent of the total regional population (including the urban centers of Choluteca, San Lorenzo, and Nacaome), and about 50 percent of the total rural population (Calculated from data in SECPLAN 1988 and 1989).

In 1988, the mean population density of the south was 72 persons per square kilometer; however, the population densities of municipalities ranged from 16 persons per square kilometer to 147 persons per square kilometer (SECPLAN 1989). The greatest concentration occurred in three areas: municipalities bordering Nicaragua to the southeast of Choluteca City; municipalities in the northern and northeastern parts of the department of Choluteca; and municipalities in the north central portion of the department of Valle. While many highland areas with the greatest concentration of rural populations corresponded to locations of earlier mining activities, lowland populations were clustered near the urban centers of Choluteca, San Lorenzo, and Nacaome. Since prehistoric times major population centers have been located in the lowland plains; the cities of Choluteca, San Lorenzo, and Nacaome, that had the largest populations in 1988, were all sites of important indigenous settlements (MacLeod 1973: 47; Stone 1957: 82–83).

Migration

The history of the south from the conquest through the 1980s encouraged a pattern of human settlement in which the greatest number of people live in the least favorable agricultural zones. Because cattle ranchers and other medium-sized landholders had controlled the lowlands since the colonial period (MacLeod 1973: 290–305; Brand 1972: 25–27), after 1880, families dislocated by

42

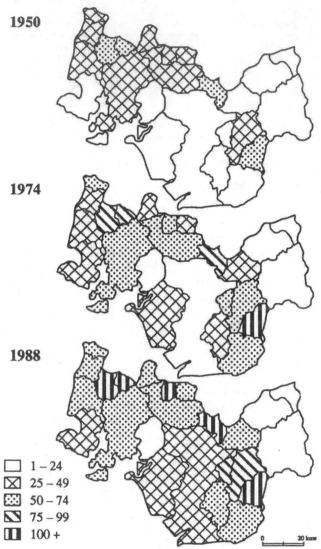

1950

1974

1988

☐ 1 – 24
⊠ 25 – 49
▥ 50 – 74
◩ 75 – 99
▥ 100 +

0   20 km

FIGURE 2.4 Population Densities of Municipalities in Southern Honduras: 1950, 1974, and 1988 (density in inhabitants per square kilometer). Source DGECH 1954, 1976. SECPLAN 1988, 1989.

the boom in export coffee production in neighboring El Salvador, Guatemala, and Nicaragua were compelled to join Honduran peasants in the southern highlands (Boyer 1982). By 1950, most of the hillsides and virtually all the valley bottom land in the highlands had been claimed. Population densities reached an average of 30 inhabitants per square kilometer with some border areas twice that. For the first time, a few peasant families faced landlessness, and by 1960, the south was Honduras' most densely populated region (Molina 1975: 31; Boyer 1991: 4).

Although earlier this century the south had attracted people from other areas of the country as well as from neighboring Nicaragua and El Salvador, by the mid-1960s this trend reversed and out-migration from the region accelerated, stimulated by mounting populations and the growing scarcity of land (Molina 1972). Migration was complex: encompassing women and men almost equally; was permanent and cyclical; included rural and urban areas as destinations; and was both intra- and inter-regional (Stonich 199a). Beginning in the 1950s, land poor and landless peasants moved to the southern lowlands looking for either land to rent or for wage work. At first, the uneven expansion of capitalist agriculture in the lowlands allowed some to find rental land in, as yet undeveloped, sections. More common, however, was migration to the growing urban centers of Choluteca, Nacaome, and San Lorenzo (Stares 1972: 80–82). The continued importance and magnitude of migration, is indicated by information from San Esteban and Oroquina, the two communities that are the focus of the ethnographic study in Chapters 5 through 7. In these villages, 39 percent of children over 13 years of age were no longer living in the communities and 76 percent of male householders and 36 percent of women householders had migrated at least once to work outside the communities.

The growth of the capitalist agricultural sector and the associated dependencies it created did not result in a corresponding expansion of wage labor opportunities. Temporary migration for wage labor most often involved the harvesting of commercial crops; especially cotton, sugar, coffee, and tobacco. In the period 1980–81, such agriculturally related jobs provided only 30 percent of the total number required (CSPE/OEA 1982: 97). Overall rural unemployment in 1980, based on the corresponding monthly supply and demand for labor, averaged 62.2 percent for the year—ranging from 15.7 percent in September to 95.5 percent in March. In reality the unemployment rate was much higher because women were not considered in the estimates (CSPE/OEA 1982: 97–99). Since then, declines in cotton and sugar cultivation, the disruption of coffee production in the highland areas bordering Nicaragua during the violence between the Contras and the Sandinistas, and the virtual closing of the Nicaraguan border to Hondurans who previously went there to participate in a number of seasonally available agricultural jobs, have decreased employment opportunities still further. Among the consequences of depressed levels of employment and the dwindling size of landholdings were extremely low levels of income. Based on the agricultural survey of 1980, per capita income in the rural sector was estimated at US$118.50 and family income at US$712.00 (CSPE/OEA 1982: 112–114).

Migrants from degraded areas in the south are also settling in the department of Olancho and the vast, relatively-unpopulated, areas of *La Mosquitia* in
northeastern Honduras. According to the national population census of 1974,
the adjacent departments of El Paraiso and Olancho ranked only after the largest cities (Tegucigalpa and San Pedro Sula) as the predominant extra-regional
destinations of migrants from the south (Stonich 1991a). Community level research indicates that by the 1980s these two departments accounted for more
than 50 percent of the total destinations of male householders in some rural
highland communities (Stonich 1991a). The first organized migration of people from the south began in the early 1970s and by the 1980s there were settled
communities along the entire upper reaches of the Rio Patuca.

The colonization of tropical forest areas has extended into the Rio Platano
Biosphere Reserve.[8] Reproducing processes occurring throughout Latin
America, deforestation has taken a heavy toll on the ecosystems, as newly-
arriving colonizers (many utilizing the illegal roads constructed by loggers)
clear forest for crops, cattle, and fuelwood, thereby facilitating the establishment of ranching interests while simultaneously encroaching on the lands in-
habited by Honduras' small remaining indigenous population.

Urban Growth

The migration of resource-poor rural family members from the south also contributed to the explosive growth of the capital city of Tegucigalpa, of San Pedro Sula, and of secondary urban centers in the south (Stonich 1991a). From
1974 to 1988 the growth in the urban population of Honduras surpassed the
overall rate of population growth. Between 1974 and 1980, the urban population growth rate was 5.8 percent exceeding the overall population growth rate
of 3.6 percent. Although the urban growth rate declined to 5.4 percent from
1980 to 1987 it continued to exceed the overall population growth rate of 3.4
percent for the same period (USAID 1989a, 1989b). Between 1950 and 1988 the
urban population of Honduras grew from 18 percent to 40 percent of the total
population—a rate of increase not exceeded by any other Central American
country (Calculated from data in World Bank, various years).

The growth of major urban centers in the south has been part of this national trend, although the south has remained relatively more rural than has
Honduras as a whole. In 1988, while the majority of people in the south (75
percent) continued to live in rural areas, 25 percent lived in or near these expanding urban centers (SECPLAN 1989). Although growth was centered in
Choluteca, it also encompassed San Lorenzo and Nacaome as secondary cities: between 1950 and 1988, the population of Choluteca increased by 772 percent, while San Lorenzo grew by 569 percent and Nacaome by 285 percent
(SECPLAN 1988).

The regional transportation system established in the 1950s linked
Choluteca, Nacaome, and San Lorenzo to each other, to the neighboring countries of Nicaragua and El Salvador, as well as to the capital city of Tegucigalpa
and the rest of the country. At that time the Honduran government became an
agent of development, creating a variety of state institutions and agencies to

expand government services and undertake infrastructural projects (White 1977). The transportation system established in the south was part of a national effort to unify the country, expand the national market, and more effectively integrate agricultural producers into world markets. The urban centers of the south were part of, and facilitated, these goals and became the sites of health care, educational, financial, and commercial activities (Stares: 1972: 80–82).

Conclusions: Spatial, Eco-Demographic, and Resource Marginality

Although southern Honduras presents serious environmental constraints and has undergone significant population growth, another critical factor must be examined in order to explain the worsening human and environmental conditions that exist there—the increasing unequal distribution of resources emanating from the expansion of export-oriented agriculture. The restructuring of agriculture in southern Honduras since World War II has impoverished both the landscape and a growing percentage of the population. The general trend has been toward resource oligopoly, patterns of exploitation and production that jeopardize future sustainability in exchange for quick profits, wanton destruction of natural resources, and underemployment. None of these processes resulted, even indirectly, from population pressure.

The region is a lucid example of a case of spatial or eco-demographic marginality in which small producers have been displaced to less fertile and more environmentally vulnerable highland locations. Although the process began shortly after the Spanish conquest, it accelerated significantly in the post–World War II period as a result of land expropriations by the state and agribusiness concerns primarily found in the lowlands. This spatial maldistribution of the population, in which the majority of smallholders are pushed into increasingly more marginal areas while a relatively small number of large producers has access to the best lands and the most resources, has advanced the devastating human and environmental conditions in the south.

An analysis of the results of the National Nutritional Study conducted in 1986, which included children attending first grade in Honduran public and private schools, indicates important associations among the percentage of chronically undernourished children, the population density of municipalities, and whether the municipality is located in the highlands or in the lowlands.[9] While the percentage of chronically undernourished children ranged from 26 percent to 47 percent among municipalities in the south, there was a significant relationship between the extent of undernutrition and the density of the municipal population. The Pearson Correlation Coefficient between the percentage of children with chronic undernutrition and the population density in the municipality was .551 (p<.004). In addition, for those municipalities classified as having "moderate" numbers of undernourished children (i.e., less than 36 percent), the average population density was 65 people per square kilometer, while those with "high" numbers of such children (i.e., 36 percent or more) had a mean population density of 85 inhabitants per square kilometer. Moreover, ten of the twelve municipalities classified as having a "high" percentage of undernourished children are situated in the highlands. This is not to say, by any means, that high population densities necessarily result in chronically undernourished children or that lowland children are adequately fed. On the contrary, the difference between the percentage of chronically undernourished children in highland municipalities (36 percent) was only slightly above that found in lowland municipalities (33.3 percent).

Neither are the severe environmental problems of the region confined to the highlands. The destruction caused by the poor in their desperate search for survival is no less alarming than the destruction brought about by large landowners through their reckless search for profit. A recent report by agricultural scientists, who have worked in the region for some years, states:

> Since the 1950s, the agricultural economy of southern Honduras has been dominated by a series of boom and bust cycles of export commodities. Cotton, sugar, and cattle each reached their zenith only to dissipate in the face of declining productivity and adverse world markets. Much of this instability has been self-inflicted through degradation of the natural resource base which has reduced productivity and profitability. At present, nontraditional export crops like melons and shrimp are experiencing the great expectations and up-swing of this cycle; however, signs of limitations and stress on production are becoming apparent (Meckenstock et al., 1991, p.2).

In southern Honduras, it is the union of greed and poverty that has produced a context in which a relatively rich few and many poor mine nature's capital. As this chapter has suggested, and as ensuing chapters demonstrate, an understanding of the relationships among population pressure, human impoverishment, and environmental degradation, is deficient without an examination of the inequalities in access to resources and other distributional considerations related to the prevailing model of economic development.

Notes

1. See CSPE/OEA (1982) for a comparison of several regional delineations. Variations in regional limits most often involve determining the northern border. They also diverge depending on whether or not they include various municipalities located in the adjacent departments of El Paraiso, Francisco Morazon, and La Paz. It should be noted that this characterization of the south as a region is not made on the basis of some artificial economic or political boundaries (although the southern region does generally correspond to the Honduran departments of Valle and Choluteca). The Honduran south as a distinctive region can be traced through numerous references in the literature from the colonial period to the present and is a spatial, social, and economic category in the minds of Hondurans throughout the country as well as in the programs and practices of national and international agencies. Government statistics frequently are broken down by region with the departments of Choluteca and Valle comprising the southern region. However, it should be borne in mind that the region is not homogeneous (e.g., obvious differences exist between the lowlands and highlands) and that some of the patterns existing in the south are easily applied to portions of surrounding departments and to portions of El Salvador and Nicaragua as well.

2. Lowland municipalities include those in which 75 percent, or more, of the municipality is found at elevations equal to, or less than, 200 meters, while highland municipalities comprise the remainder: i.e., less than 75 percent of the municipality lies below 200 meters. Assigning municipalities to the two groups was done in consultation with staff from the office of the Comprehensive Resource Inventory and Evaluation System Project, Michigan State University, on the basis of data found in CRIES (1984) and unpublished data. According to this scheme, sixteen municipalities are defined as highland: Concepción de Mariá, Oroquina, Langue, San Antonio de Flores, San Francis de Coray, El Corpus, Pespire, San José, San Isidro, Caridad, El Triunfo, Goascorán, Aramecina, Apacilagua, Duyure, and Morolica. The nine lowland municipalities include: Yusguare, Amapala, Namisigue, Nacaome, Alianza, Marcovia, Choluteca, San Lorenzo and San Marcos de Colón.

3. As will be discussed in Chapter 4, production of melons for export began in the 1970s, but not at the scale of the 1980s operations.

4. Dominant mangrove species are *Rhizophora mangle, Rhizophora samoensis, Conocarpus erecia, Laguncularia racemosa, Avicennia bicolor*, and *Avicennia gerinans*.

5. Important commercial shrimp species include *Penaeus occidentalis, Penaeus vannamei*, and *Penaeus stylirosiris*.

6. Using regional boundaries suggested by the Honduran National Economic Planning Council (at that time CONSUPLANE) and the United States Department of Agriculture (USDA) soil suitability classes, Schreiner and Badger (1983) in their proposed regional development plan for the south concluded that 10 percent of the soils in the region are in Classes I to III (i.e., suitable for intensive cropping). A further 20 percent is in Class IV (suitable for pasture). Some of the remaining 70 percent of the land classified as Class V to Class VIII is also suitable for pasture. Such evaluations ignore both the reality and the human necessity of the intensive agricultural production that takes place in the areas judged as "most unsuitable" for that type of cultivation. This is further illustrated by Schreiner and Badger's broad classification of soils in the south as either infertile or relatively fertile (due to slope, drainage, and rainfall) (Schreiner and Badger 1983: 14).

Problems of inconsistencies and overly broad classifications were considered by the Comprehensive Resource Inventory Evaluation System Project (CRIES) in their natural and agricultural resource assessment of the department of Choluteca (CRIES 1984). Using the descriptive classification materials already available supplemented with limited field research, CRIES reclassified southern soils in terms of a common system (USDA Soil Taxonomy). For those areas for which no pedological classification was available classification was inferred from available data on geology, climate, vegetation, and topography. Specific soils information was obtained from soil maps (scale 1/10,000 and 1/20,000) prepared by the national cadastral agency and USAID. Soil units were defined by one or two soil series and by slope ranges. (Soil unit descriptions, area totals, and percentage distribution of Choluteca soil units are found in CRIES 1984: 32–40.) Soil properties within particular soil units which affect crop production such as texture, natural drainage, depth to bedrock, Ph, salinity, and slope were assessed. On the basis of their analysis the soils of Choluteca vary from sandy to clayey, from shallow to deep, from well to poorly drained, and from nearly level to very steeply sloped. At a depth of 50 centimeters, mean annual soil temperatures are 22 degrees Centigrade or higher throughout most of the department but can range from 15 degrees Centigrade to 22 degrees Centigrade at higher elevations. Mean soil temperatures vary less than 5 degrees between the rainy and dry seasons. The soil moisture regimes found in the department of Choluteca are aquic, ustic, and udic. The aquic moisture regimes are saturated with water, virtually free of dissolved oxygen, and may require artificial drainage for good crop production. Soils with an ustic moisture regime have limited moisture but sufficient for plant growth during a limited portion of the year. They may require supplemental irrigation for certain crops. Udic soil regimes are not dry for any period of at least 90 days and have a mean annual soil temperature of less than 20 degrees Centigrade. They are commonly found in the more humid climates where precipitation is evenly distributed during the year.

To evaluate soil units best suited for specific crops an assessment of production potential was made by CRIES which compared the physical properties of each soil unit defined with the physical requirements of primary crops. Limited fieldwork was conducted and discussions with Honduran counterparts having expertise in soil science and agronomy were also held to aid in making the assessments. Each crop was rated as having either a low, low to medium, medium, medium to high, or high potential for production in each soil unit. Although elevation was not a criteria to define soil units, particular units were generally associated with particular ranges in elevation. Keeping this is mind, it was possible to spatially locate soil units and characterize them as either highland or lowland soils. By and large highland soil units were rated as having low production potential for almost all crops, while lowland soil units were rated predominantly as having medium, medium-high, and high potential. However, as will be discussed in subsequent chapters it is in areas where these highland soil units occur that the most intensive agricultural productions occurs, whereas lowland areas dominated by soil units having the highest potential are the sites of commercial agricultural production.

7. Durham (1979: 123–124) estimated that in 1969, 300,000 Salvadorans (12 percent of the total Honduran population) were living in Honduras, of whom about half had arrived after the beginning of the boom in commercial agriculture. Of these, approximately 80,000 were small farmers (Durham 1979: 126).

8. The Rio Platano Biosphere Reserve located within the region referred to as *la Mosquitia* was established in 1979 and became a world heritage site in 1980. It covers almost the entire watershed of the Rio Platano, approximately 525,000 hectares on the Caribbean coast of Honduras. It the time it was created there were approximately 4,450 inhabitants, mostly Miskito Indians and a few Pech and *ladino* villages within the boundaries (Poole 1989). At that time the reserve was inaccessible by road. By 1990 there were still no conventional roads into the reserve, however, illegal access was facilitated by widespread logging and gold-mining. By then, several hundred Nicaraguan Miskito Indians had settled within the reserve.

9. This study was done by the Honduran Secretary of Public Education and the Institute of Nutrition of Central America and Panama (INCAP) in 1986. Details can be found in SAEH/INCAP (1987). Height, age, and sex data were collected from 80.1 percent of registered first graders between the ages of 6 years, 0 months and 9 years, 11 months attending public and private schools in Honduras. Students who were more than two standard deviations below the height for age ratio of the WHO reference population were classified as chronically undernourished. Students who fell between 2–3 standard deviations below were classified as being moderately undernourished; students who were less than 3 standard deviations below the reference population were classified as severely undernourished. At the national level, 39.8 percent of first graders were classified as chronically undernourished—27.6 percent with moderate undernutrition and 12.2 percent with severe undernutrition. For southern Honduras, 35 percent of children were determined to be chronically undernourished—26 percent with moderate undernutrition and 8 percent with severe undernutrition.

Municipalities with 20–35.99 percent of chronically undernourished first graders were classed as having moderate levels of undernutrition while those municipalities with more than 36 percent of such children were classified as having high levels of undernutrition.

3

The Historical Legacy

The sequence of boom and bust cycles in export commodities that occurred in southern Honduras since the end of World War II, is the most recent manifestation of a long history of uneven development.[1] Since the Spanish conquest, the south underwent repeated attempts to develop the region and recurring periods of economic expansion and contraction. Until the post-war period, however, the various efforts to promote development and incorporate the region into the national economy were erratic and incomplete. Honduras generally is considered to have become articulated into the world economic system at the end of the 19th century as a result of the externally financed establishment of the banana industry and the reintroduction of mining. The complete integration of southern Honduras came later—only after 1950, with direct national and international ventures to foster economic development in the region. However, the socioeconomic, demographic, ecological, and agricultural conditions that existed in the south at the time of more complete integration, were descended, in no small measure, from patterns established during earlier periods of partial integration.

Southern Honduras Before the Spanish Conquest

The Spanish conquest had catastrophic social and economic consequences on the region's indigenous peoples and caused their almost complete annihilation. Nevertheless, the bases of contemporary patterns in small-scale agriculture, and other kinds of artisanal production dependent on the region's natural resources, are found in the prehispanic past. Archeological evidence from southern Honduras is sparse and poorly understood, and except for stratigraphic excavations by Baudez (1966), the south remains largely unexplored.[2] Apparently, the south was not integrated strongly either internally or externally with highland Mesoamerican or lowland South American cultures. It was rather a confluence of northern (Mexican), southern (lowland South American), and indigenous Central American peoples (Stone 1957: 82–84; MacLeod 1973: 26–46). At the time of the conquest, the Chorotega-Mangue appear to have been dominant followed in number by the Ulva and finally by the Poton (Stone 1957: 83).[3]

Most groups practiced slash and burn agriculture, cultivating corn, beans, and squash. Although these made up the largest part of their diet, these staples were supplemented by chiles, peanuts, cacao, fruit, turkeys, and dogs (Stone 1957: 86–87). Equally important was the *monte*, uncultivated land where people gathered nuts, roots, and grubs for food, as well as hunted and trapped deer, iguana, birds, and jaguars for meat and skins. The *monte* also served to supply brush, timber, and reeds for construction (MacLeod 1973: 215). Along the coast, the Poton collected shellfish and fished the coast in dugout canoes (Stone 1957: 86). Cacao and cotton were cultivated as well (Stone 1957: 86) and it is likely that the region participated in coastal and hinterland trading networks.

The Periods of Conquest and Colonization

Sixteenth century chroniclers reported large, settled, indigenous populations in most of Central America, particularly on the Pacific Coast with its rich, volcanic soils which stretched from Guatemala to Panama (see Denevan 1976 and 1992 for regional population assessments). Estimates of the preconquest population of Honduras range from 400,000 to 1,200,000 people.[4] By 1540, when Benzoni visited Honduras, he reported that only 8,000 native people were left alive. According to MacLeod, by 1540 the Chorotega and other indigenous groups had "ceased to be nations and had become linguistic classifications" (1973: 100). Ethnohistoric evidence suggests that by 1582 the entire Pacific Coast of Honduras may have been reduced to less than 750 families (Brand 1972: 12).

The precipitous decline in the indigenous population was due less to military operations at the time of conquest than to the more far reaching consequences of malnutrition from disruptions of food production and distribution systems, a series of local and pandemics, slave trading, and the use of indigenous people as slave labor in mining and as bearers. Several regional and pandemics, with mortality rates of from one-third to two-thirds of the indigenous population, were reported in Central America as early as 1519, and in Honduras, as early as the 1520s. These epidemics continued through the next two centuries and by most accounts affected the indigenous peoples more severely than the Spanish because of the vulnerability of their immune systems to European diseases and their weakened physical condition (MacLeod 1973: 98–100).

What Spain sought in the New World was wealth. First gold and then silver were mined and sent to Spain as bullion to underwrite Spain's military operations in Europe, support an unwieldy growing bureaucracy, and settle European debts. Honduras was among the places that the first conquerors found gold and silver, and perceived as a potential source of great wealth, it was a center of early Spanish activities. The rapid depopulation of southern Honduras occurred as a consequence; mining activities were begun quickly and between 1520 and 1550 large groups of people were taken as slaves to work in the mines of Honduras or sent in slave ships to Panama and Peru (MacLeod 1973: 54–56; Wortman 1982: 3–7). Wortman, describes the period:

By far the most destructive search for wealth occurred in Nicaragua and in Southern Honduras. The slavers sought bodies for export. They raided and depopulated areas that were said to have had "more people than hairs on all the deer."
The commerce continued until mid-century when depopulation, royal legisla-
tion, and the lure of Incan wealth to the south put an end to it leaving a vacuum
for other populations and for cattle to fill (1982, 5).

By the third quarter of the 16th century, the Spanish conquest fundamentally had altered demographic, social, economic, and land use patterns in
southern Honduras. The lowlands were virtually depopulated and the few indigenous peoples left were reduced to remnant populations. In the highlands,
relatively more indigenous peoples survived and were joined by others who
fled the lowlands. Although the Spaniards established permanent settlements
in the lowlands, first near the present city of Choluteca in 1522 and somewhat
later in Goascorán and Nacaome, these settlements remained small, composed of only a few families (Stone 1957: 82–85).

This economic and settlement pattern changed in the south, after major silver strikes were made in the highlands around Goascorán and Tegucigalpa beginning in 1569 (Stone 1957: 82–83; Brand 1972: 18–19; MacLeod 1973: 148–
149). Of a much larger scale than the first attempts at gold panning and silver
mining in the region, these strikes altered economic life throughout Honduras.
By 1580, thirty small silver mines were operating in the central highlands and
silver production rose rapidly before reaching a peak in 1584. The center of
Honduran economic activity moved from the Caribbean coast to the central
highlands and resulted in the initial Spanish settlement of the area around the
present capital of Tegucigalpa. Heretofore, unsubdued highland indigenous
peoples, served as labor for transportation and smelting and as providers of
food when the *encomienda* system began (MacLeod 1973: 148). The mining
complex required regular supplies of meat, leather sacks, tallow and mining
candles, and stimulated the development of stock raising in the south and central regions. Both cattle and mules were raised on the plains around Choluteca
and along the Gulf of Fonseca. Hides and meat from cattle were shipped to
mining areas, and mule trains were sent up to the mines in the highlands or to
Panama. Salt, needed in the smelting process, encouraged the Spanish take
over of the indigenous salt-making industry along the shores of the Gulf of
Fonseca (MacLeod 1973: 253–263). Faced with a constant labor shortage
caused by the initial depopulation, many highland indigenous peoples, as
well as many remaining lowland peoples, were placed in *reducciones*, small
settlements located predominantly in the highlands near mining camps where
they served as a source of labor (West 1959: 767–777).

Despite these far reaching effects on indigenous people and on the land,
Honduran mining remained only marginally profitable. After reaching peak
production in 1584 it began to decline and from then through most of the 17th
century, the industry was wrought with fraud, deficient in capital and labor,
ignored by Spain, and unable to solve the technical problems posed by the difficult local terrain (MacLeod 1973: 148–151).

One further attempt was made to economically develop the south in the latter 16th century. Between 1590 and 1620, much of Central America's Pacific Coast, including the lowlands around Choluteca, became the site of indigo plantations. Hardier, and easier to care for, indigo began to replace cacao in many areas along the coastal plain, especially in those places where earlier attempts to produce cacao for export had met with little or marginal success. Labor requirements limited to two months during the harvest and dye extraction process, made indigo seem well suited to times of economic depression and depopulation. Despite this potential, high transportation costs and the lack of convenient markets stifled the industry, which stagnated from the 1620s until the end of the seventeenth century (MacLeod 1973: 202–203).

Its rugged terrain, which made transportation from the Pacific to Caribbean coast extremely difficult and costly, hindered the political and social integration of Honduras and inhibited the early development of profitable export industries, and thus the establishment of a unified national economy. Southern Honduras, with its emphasis on livestock, mining, and later indigo production, was but one of a collection of separate regional economies within the country. It remained a backwater within a Honduras that was itself geographically, economically, and politically peripheral to the Spanish American empire. This marginalization continued during the general economic depression of the Spanish empire after 1650 and the economy of all of Honduras stagnated through the 18th century (Woodward 1976: 75).

The legacies of the Spanish conquest and colonization included not only social, political, and economic fragmentation, but also the start of existing inequalities (Villanueva 1968; Perez Brignoli 1973a, 1973b). The small privileged minority in Honduras were rural elites whose wealth was concentrated in the private ownership of large tracts of land. From early colonial times, private ownership of very large amounts of land were granted to knights, captains, and squires of the court, while much smaller grants were given to Spanish soldiers and *peones* (Villanueva 1968). A comparison of early land titles registered between 1600 and 1649 shows that of the 47 total grants, 42 (89.4 percent) were given as private property (Villanueva 1968: 66). Large grants included at least eight in the lowland south: a grant to the Franciscans to establish a monastery in Nacaome (Stone 1957: 100); a concession of 16 *caballerías* (721 hectares) to the Mercedarian monastic order in 1607 (MacLeod 1973: 302); and six other grants near Choluteca large enough to accommodate a total of approximately 30,000 cattle (MacLeod 1973: 304). Private land grants continued to constitute more than eighty per cent of all recognized land claims during the colonial period (Villanueva 1968: 66).

Generally, the largest *haciendas* were located in the lowlands and were characterized by individual, or family control, of large tracts of land. Income was usually derived from ranching and the labor of peasants and indigenous peoples under a variety of tenure arrangements; debt peonage, sharecropping, or tenant farming (MacLeod 1973: 290). Medium sized holdings, found both in the lowlands and in some highland valleys were much smaller, averaging about four *caballerías* (180 hectares). On these, large Spanish and mixed families lived modest lives, raised a few cattle, and cultivated a few plots of corn and perhaps some sugar and indigo (MacLeod 1973: 302). Despite differences in the size of land holdings, labor shortages, trade restrictions, and the lack of transportation facilities inhibited capital accumulation for the largest landowners whose standard of living and way of life was much the same as that of their medium land owning neighbors (MacLeod 1973: 302–305). The proliferation of private grants during the colonial period, however, did not eliminate the existence of communal and ejidal land; during that time, indigenous and ladino groups in the south successfully petitioned for common or municipal land (MacLeod 1973: 222).

Post-Colonial Nineteenth Century and Early Twentieth Century

The considerable intellectual, economic, and political changes that characterized the century preceding independence in Guatemala, El Salvador, and Costa Rica, were not matched in Honduras. The mechanisms of economic integration binding communities, market centers, and regions that developed during colonial times in those countries were virtually absent in Honduras. In Guatemala, El Salvador, and Costa Rica, the decline of Spanish political hegemony and the rise of British commercial dominance in the region, provided the context for the emergence of a middle class and for changing leadership among the creole aristocracy which in turn set the stage for the political power struggles that were to come. Political parties having fundamental ideological and economic differences struggled for control in a framework of declining indigo markets—the most important export commodity of the time. At independence, as during most of the colonial period, Guatemala remained the region's center of power, while the smaller, weaker, and poorer provinces of Honduras and Nicaragua feared domination by their stronger neighbors (Woodward 1976: 89–91; LaFeber 1983: 25–28).

Given the peripheral position of Honduras within the Spanish colonial, mercantile system, the relative weakness of its elites, and its unarticulated national social, political, and economic systems, it is no wonder that the early decades of Honduran independence were characterized by extreme political instability; between 1826 and 1876, eighty different governments were inaugurated. The relative weakness of the Honduran political structure also made it an easy target for neighboring Liberal and Conservative governments in Guatemala and El Salvador that sought to consolidate their positions at home and to extend their influence by establishing sympathetic regimes in Honduras. Within Honduras, local elites, retaining greater wealth in cattle than in other commercial enterprises, sought political control at the municipal and national levels through *caudillo* politics (Brand 1972: 42–43).

In the late 19th century, following the collapse of the Central American Republic, the Liberal reforms that promoted export economies in Guatemala, El Salvador, and Costa Rica, corresponded to the consolidation of national power by agrarian oligarchies that emerged primarily through the dramatic growth of coffee export economies. Facilitated by British capital, these reforms confirmed the declining influence of the conservative colonial and religious aristocracy and provided the basis for the incorporation of these countries into the international economic system (Torres Rivas 1973: 59–73). Such efforts were constrained in Honduras by the political, social, economic, and regional isolation already mentioned. Nevertheless, through the intervention of the Guatemalan government under President Justo Rufino Bárrios, the regime of Aurelio Soto began carrying out Liberal reform in Honduras in 1876 (Stokes 1950: 42). However, the promotion of export agriculture and limited monetary and fiscal reforms, did not signify a new balance of social, political, and economic power in Honduras as they did elsewhere in Central America (Perez Brignoli 1973b: 42–43). Consequently, when foreign capital successfully was sought as a stimulus to domestic industry, it did not supplement or compete with an established export economy (Finney 1973: 26). These conditions made possible the immense power of the foreign owned mining and banana companies, as well as maintained the continued relative weakness of the Honduran elite through the first half of the twentieth century.

Mines and Bananas

Regardless of Honduras' historically marginal position within the wider world system, it became irretrievably bound to it with the investment and expansion of foreign owned mining and banana companies in the 1880s. Although the role of banana companies in Honduras is better known, it was the introduction of the mining companies that established the pattern of foreign company investment and behavior in Honduras. Believing that a revitalized mining industry would entail widespread social and economic benefits for the country as a whole, a succession of Liberal governments sought to induce taxable foreign investment in the country. As an added incentive, a variety of concessions were offered to those willing to invest; land grants, tax exemptions, and rights to other natural resources such as timber, water, and limestone. The result was an influx of fortune hunters and nearly 100 mining companies were organized in Honduras between 1880 and 1900 (Finney 1973: 25–32). Most of these enterprises failed quickly due to lack of capital, and only the New York and Honduras Rosario Mining Company were moderately profitable for any length of time (Finney 1973: Chapter 1).

Nevertheless, these companies did have significant, although unforseen, social, economic, and political effects. The capital intensive nature of mining precluded the employment of a large labor force and while native Hondurans were trained for the limited number of unskilled and most dangerous jobs, the few skilled jobs were held by foreign workers who had been brought into the country by the mining companies (Finney 1973: 260). Although the relatively high wages paid to miners increased the demand for consumer goods, the limited extent of this demand favored imports over the development of domestic production. While the boom provided modest stimulation to service and supporting industries such as livestock, food crops, wood and salt, the growth of these activities was dependent on the continuing expansion of mining which did not happen (Finney 1973: 330–363). The main beneficiaries of the mining boom were those who had encouraged it—the Honduran national authorities.

The companies did serve as a source of credit and did modernize the fiscal and monetary systems of the central government in addition to distributing personal favors among top Honduran officials (Finney 1973: 330–363). In return, the national authorities usually supported the companies in their disputes with local communities over questions of property, labor and legal rights (Finney 1973: 338, 392–393). Thus, a major consequence of the mining boom was to centralize the heretofore fragmentary power of the national government. The cost was the initial alienation of Honduran control over their national economy. In the fiscal year 1887–88, precious metals comprised 52.2 percent of the value of Honduran exports of which the Rosario Mining Company's share was 86.6 percent (Perez Brignoli 1973b: 15–16). Although the mining boom ended by the turn of the century, the pattern of interaction between foreign companies, Honduran officials, and the people of Honduras was well established by that time.

Between 1880 and 1913, this pattern became more firmly entrenched with the initial investment and expansion of U.S. capital to produce and export bananas.[5] The far-reaching effects of this investment are well beyond the scope of this volume; however, several important consequences are relevant: (1) It reinforced the earlier development in mining by linking Honduras more closely to the world economic system as an exporter of primary commodities (henceforth derived from tropical agriculture) and an importer of manufactured goods; (2) Through virtual monopoly by U.S. companies, United Fruit, Standard Fruit, and Cuyamel Fruit (which became part of United Fruit in 1929), it subordinated the Honduran economy to decisions and forces beyond national control; (3) Together with the mining boom, it fostered the growth of a strong central government and bureaucracy in the midst of a weakly integrated society; and (4) With the concentration of the banana industry to the north coast, the response of regional elites in other parts of Honduras was to focus their own economic activities in the highland interior and on the Pacific Coast.

The Effects in Southern Honduras

In the south, the colonial patterns of sparse settlement and of scarcely differentiated large and medium landowners in the lowlands persisted until at least the mid 19th century after which some economic, demographic, and social changes seem to have occurred. In the 1880s, as part of the Liberal effort encouraging foreign investment, 20 percent of all concessions granted by the national governments were in the south (Brand 1972: 100). During this period, British and U.S. companies re-opened colonial mines or opened new mines in other central and southern highland areas. Mining companies rarely operated their own company stores and as a result had to rely on local producers and merchants. Highland peasant farmers in many areas began producing for the market; supplying mining complexes with corn, beans, coffee, cacao, sugar cane, and fruit (Finney 1973: 335). Cattle ranchers along the coast and in the highland valleys diverted portions of their cattle to the mining camps (Finney 1973: 348). Local merchants in the regional centers of Choluteca, Nacaome, and Aramecina expanded product lines and total sales in order to satisfy the needs and desires of the relatively highly paid mine workers and their families (Finney 1973: 334). Infrastructural support of these commercial activities included the building of wagon roads between the mines, the port of San Lorenzo (which became part of the regular shipping lanes to San Francisco), and the capital city of Tegucigalpa (Brand 1972: 100–103, 156).

Mining and related activities attracted not only Hondurans from outside the region looking for work, but also Salvadoran, Welsh, Italian, and German families. After the mining boom ended, some of these families remained and became members of an emerging merchant elite in the small urban centers of Choluteca and Nacaome and other towns in the south (Boyer 1982: 66). During the period 1889 to 1901, the population in the south grew from 10 percent to 14.5 percent of the total national population, reflecting the in-migration to the area (Molina 1975: 60; DGECH 1981: 47) and reached a regional population density of 13.7 inhabitants per square kilometer by the end of the period (Boyer 1982: 69).

In 1893, the present day political divisions of the south were demarcated with the creation of the department of Valle in the western portion of the region adjacent to El Salvador. Valle along with the department of Choluteca which was created as part of the initial partition of the Honduran territory after independence in 1825 (Rubio 1953: 69–70), have remained the two major political subdivisions of the southern region. The census of 1901 shows Choluteca with 16 municipalities and Valle with eight. Commonly, each municipality contained a small municipal urban center rarely of more than 150–200 people and scattered outlying ranches and homesteads where the bulk of the population lived (Boyer 1982: 69).

Thus by 1900, settlement in the south consisted primarily of large and medium sized cattle ranches in the lowlands and scattered peasant homesteads in the highlands. After independence, the Honduran government accelerated a process begun in colonial times and continued to make an increasing number of *ejidal* grants to communities. These served to protect and to guarantee access to land to small farmers to such an extent that through the 1880s, "each department had sufficient agricultural land and each *campesino* was able to have the land necessary to support himself and his family" (Carías and Slutzky 1971: 34). In the urban centers, a small foreign dominated merchant class was beginning to form. By that time also, ladinoization had occurred to the extent that there were few indigenous peoples left. Those who remained lived primarily in the remnants of the *reducciones*.

Contemporary patterns in the location and establishment of peasant rural communities also began to be formed during this period, although to a great extent, the highlands remained sparsely settled. Many of the areas of greatest population concentration in the highlands were the sites of colonial mines or mines started during this period (Finney 1973: 338). Many of the first rural communities were settled by families from outside the region; from northern Honduras, Nicaragua, and El Salvador. Most often, these outlying settlements consisted of six to ten households each comprising two to four extended families scattered over a contiguous area of mountainous terrain. The first families to arrive in an area, usually occupied from 70 to 100 hectares of land on which they cultivated small plots of corn, sorghum, beans and other subsistence crops, and raised a few cattle. Most production was for home consumption with household surpluses first distributed to extended family members and any excess carried to the nearest municipal center for sale. Initial families usually had access to more land than did subsequent settlers, and consequently tended to become dominant in the economic, political, and religious affairs of the community. The family patriarchs often became community leaders and represented the community in local politics by giving loyalty to municipal Liberal or National party bosses in exchange for legal and protective favors (Boyer 1982: 70; field interviews).

More Complete Integration into the National, Central American, and World Systems: The 1930s to the 1950s

The economic and political aspirations of Liberal reformers had been stifled by nine major civil wars in the period from 1893 to 1932, by the worldwide economic depression during the 1930s, and finally by the conservative, repressive regime of Carías (1932–1949). A concomitant to Carías' dictatorship was an end to the violent political wars and a more politically and socially integrated nation. In brief, Carías ruled by giving the U.S. fruit companies almost complete freedom on the north coast while simultaneously discouraging major capitalist and infrastructural expansion elsewhere (Marinas Otero 1963: 133). Already having incurred heavy foreign debt, Carías curtailed the establishment of transportation and communication systems in the rest of the country for fear of increasing this debt further (White 1977: 93).

Despite Carías' reluctance, important changes arose in the south during his regime. Although subsistence production continued to predominate in highland peasant communities, production for the market expanded, due to increased opportunities brought about by growth in the size and number of settlements and communities—the result of natural population growth and in-migration to the region—and by the existence of urban merchants. In addition, property relations became more formal—land titles registered with municipal authorities began to replace the informal land claims made by the initial families. Very clearly, a more rigid, hierarchical, rural socioeconomic status system was emerging in the highlands. Incoming families had access to less land than had the first settlers. Differential access to land began to be associated with differential access to the labor of family and friends, and this disparity in status was expressed through combinations of political and religious influences. The hierarchical status system emerging in the highlands was reinforced by Carías who tied traditional *caudillos* at the community, municipal, and regional levels into the hierarchical structure of the Nationalist Party (White 1977: 93).

Despite these changes in highland settlement, economic, and social structure, southern Honduras remained largely economically disarticulated from the rest of the nation and in turn from the world. In the lowlands, the traditional rural elites, far wealthier than any of the emerging highland elites, were still most often socially indistinct from their medium land owning farmer neighbors. More thorough entrance of the south into the world system came only at mid-century and coincided with changes in both the international economic system and the political and economic systems within Honduras.

To understand the social and economic changes that occurred in the south after mid-century, it is necessary to understand the international and national contexts in which they occurred, for changes in the south were very much a part of these larger processes. During World War II, motivated by security considerations, the U.S. government began to provide aid to Honduras in the form of road building and health assistance as well as in military training (White 1977: 97–98). These initial improvements in transportation were not focused in the south but rather linked the coffee growing areas of western and central Honduras to the north coast, San Pedro Sula, and Tegucigalpa (Stokes 1950: 12). U.S. government activities reinforced changes that had already begun in the production strategies of the fruit companies. In the 1930s and early 1940s, banana production fell dramatically due to the world wide depression, World War II, and disease. By 1940, the major banana companies began to diversify; they expanded their exports to include livestock, lumber, citrus fruit, and coconut, and also began producing for the domestic market. Expanded production included cattle raised for milk and meat (U.S. Dept. of Commerce 1956: 194–200). From then on, their involvement in both foreign and domestic markets gave the companies a vested interest in the growth of both Honduran exports *and* the domestic economy. These developments within Honduras, along with the post-war expansion of world markets for agricultural commodities, and the accompanying sharp price increases made Honduras an increasingly attractive location to foreign investors and to Honduran nationals.

Thus, when Carías' chosen successor, Juan Manuel Gálvez, took office in 1949, increased economic development had already begun on the north coast. This growth was based not only on the traditional export sector but also on the export of coffee and production for domestic consumption of a variety of agricultural products. With a transportation system already in place in central Honduras and on the north coast, trade increased quickly and with it a demand for immediate improvements in government services, particularly in the area of Honduras' primitive monetary and fiscal policy (White 1977: 94–95). Before 1950, Honduras lacked a central bank, and the prevailing medium of exchange was U.S. currency which entered through the fruit companies (Marinas Otero 1963: 126–131). Shortly after his inauguration, Gálvez received a mission from the newly established IMF which had chosen to work in Honduras precisely because of the chaos in its financial system. With their help, his administration created the Central Bank and the National Development Bank in 1950; the former having control over monetary matters, the latter providing credit for specific development projects (White 1977: 95). Shortly, Honduras became a target for an expanding network of agencies concerned with supporting expanded government activities, services, and development including: U.S. and U.N. agencies, The Economic Commission for Latin America (ECLA), international labor federations, and overseas aid funds of various religious organizations (White 1977: 94–100). As a result, Honduran government revenues and expenditures increased three fold between 1945 and 1955 (U.S. Dept of Commerce 1956: 205). During the same period, capital expenditures (spending on development) grew more important with the bulk of funds going toward road construction and to the national development bank (ECLA 1955: 117).

The Impact on Southern Honduras

As part of the government's efforts to stimulate national economic growth, the south for the first time was drawn intimately into national, Central American, and foreign (especially U.S.) markets. During the 1950s and early 1960s, the U.S., The World Bank, and IDB helped fund a variety of projects in the region. Port facilities at Amapala and San Lorenzo were improved. In 1959, the Pan American Highway linking Nicaragua and El Salvador with a section to Tegucigalpa was completed. The Pan American Highway passed through the south's major urban centers of Choluteca and Nacaome which were linked to other municipal centers and to the hinterlands by a system of penetration roads. Government supervised bus routes were established and private buses began providing services for peasants and merchants (Boyer 1982: 90).

This period of intensified public sector efforts, coincided with temporarily high prices for primary commodities on the world market. Large landowners in the south historically had been unable to respond to such potential because of the lack of necessary infrastructure, markets, and credit. With these in place, owners found it profitable to expand production for the export market. Although the terms of trade later became unfavorable, by that time much of the agrarian economy of the south had been transformed.

Conclusions: The History of Uneven Development

European colonizers very quickly established monopolistic control over the most productive lands in the lowlands, which led to the formation of large landowning rural elites. The needs of the primarily lowland hacienda owners and highland mine owners for cheap labor was satisfied through the establishment of feudal like bonds over the small number of surviving indigenous peoples who became quickly ladinoized. Integration of the region into larger economies was not consistent, and expansion did not take place at an even pace. Rather, cycles of expansion and stagnation brought periodic crises in production that necessitated the restructuring of productive forces. Nevertheless, because of low population densities, there appears to have been sufficient land available for rural peoples to retreat to subsistence production during the recurring bust cycles in mining or agriculture. Moreover this period saw the establishment of land settlement and contrasting highland and lowland agricultural patterns that persist until today. Major contemporary economic constraints and incentives rooted in regional patterns of land distribution and in land allocation emerged during this period. The long time depth involved in the development of current patterns suggests the degree to which these patterns are entrenched in the south.

The uneven nature of development in the region is shown not only over time, through the recurring periods of expansion and stagnation (boom and bust cycles), but also over space and society. The region, itself was peripheral to the nation, which was, in turn, marginal to the larger Spanish empire. These relationships persisted during the colonial and post-colonial periods. After independence, the region remained marginally integrated into the national system. Within the region, differential development occurred between the highlands and the lowlands with all of the social, economic, and demographic consequences mentioned above. For certain groups of people, these early attempts at development were both uneven and destructive. The process of incorporation into the Spanish empire devastated the indigenous population, and bound the few remaining peoples in feudal and sometimes slave like ties to the emergent elites for whom they served as a labor pool. During the many bust cycles, this increasingly ladinoized labor pool returned to subsistence production which was possible because adequate amounts of land for their use still existed in the highlands. The persistence of this long-established relationship between elites and labor until the present day, is discussed in later chapters; however, on the basis of the history of this earlier period, one can already begin to see widespread ecological changes and the costs of development for a large segment of the population of the south.

Notes

1. The concept of the "uneven" nature of capitalist development implies that the development process simultaneously has beneficial consequences for some nations, peoples, regions, etc. and negative outcomes for others, and that, therefore, development and so-called underdevelopment are part of the same process (Adams 1970; de Janvry 1981). Uneven development can occur in a number of dimensions: geographically or spatially, in the exploitative relationships that hold between developed nations and underdeveloped areas or within such areas; socially and economically in growing disparity between returns to those with power and to the rest of society (or segments of societies); and temporally, in recurring periods of growth and stagnation (Adams 1970, 1975).

De Janvry (1981) maintains that capitalist development is both uneven and integrating; integrating in that it ultimately forms an interrelated world system and uneven because it is not homogeneous, linear, and continuous, but rather is marked by inequalities over time, space, and individuals. Historically, development has been characterized by sequences of periods of expansion and stagnation (i.e., growth and recession), that have resulted in the accumulation of capital. According to de Janvry, it is in overcoming the obstacles to growth that new growth is created while at the same time new obstacles emerge (stagnation and recession). Spatially, development in particular locales, regions, or countries, necessarily is associated with retarded or deformed development in other areas (de Janvry 1981: 1–3). For individuals, capital accumulation related to development is associated with socioeconomic differentiation into classes and by class transformations. A necessary implication of uneven development is the intermittent succession of periods of reformation that are at once evidence of unevenness and the mechanisms by which the capitalist system readjusts its component parts, thus erratically but continuously transforming itself. De Janvry defines these periods of systemic regulation as two types of "crises": objective crises of accumulation manifested by economic stagnation, inflation, recession, and sharply uneven development in farms, crops, and regions; and subjective crises of legitimation characterized by massive rural poverty, increased questioning of existing social relations, political instability, and enhanced class struggle (1981: 2). Using these criteria, southern Honduras has been undergoing such crises for some time.

2. See Stone (1957) for a general overview that incorporates both archaeological and more plentiful ethnohistoric evidence. Note, however, that her overview was done more than 30 years ago using earlier analytical frameworks that emphasized migration of large populations over indigenous development.

3. The Chorotega are thought to have been of Mexican origin, having migrated from Chiapas in the fourth century A.D. to settle around the Gulf of Fonseca, Nicaragua, and Nicoya (Chapmán 1960: 94–95; Stone 1957: 82–83). Stone believed that the Ulva were native to Central American and arrived on the Pacific Coast even earlier, settling in eastern El Salvador, in the eastern portion of Choluteca, and in western Nicaragua (Stone 1957: 84). Stone accepted the Poton, who lived both on the islands in the Bay of Fonseca and on the mainland in the Choluteca region, as a Lencan group of probable South American origin (Stone 1957: 56; MacLeod 1973: 27, 33).

4. Benzoni (1967:31) estimated the population to be 400,000. Johannessen (1963:31) assumed these to be tributaries (referring to heads of households) and multiplied this figure by three (the assumed average number of people per household) to arrive at his estimate of 1,200,000. Although at first glance Johannessen's estimate may seem high, it appears more reasonable in light of the reports of Las Casas (1957–58: Volume 5:146) and Oveido (1959: Volume 4:385) both of whom estimated that from 400,000 to 500,000 indigenous people from neighboring Nicaragua were sent as slaves to Panama and Peru, another 60,000 were killed in battle, and approximately 10,000 were baptized.

5. The literature on the role of banana companies in Honduras is extensive. For this early period see especially Keppner and Soothill 1935; Keppner 1936; Wilson 1947; LaBarge 1959; Brand 1972; and Lainez and Meza 1973.

4

Agricultural Development in Southern Honduras: The Human and Environmental Effects

The evolving political economy of the south since the Spanish conquest, set the stage for the transformation which took place after World War II. The pattern of human settlement, in which the highlands became the site of more dense populations while the lowlands were more thinly settled and dominated by a small number of large and medium sized ranchers, was established shortly after the conquest and persisted for centuries. To a certain extent, so did lasting patterns of economic dependence on natural resource based products (agricultural commodities and minerals), the control of those resources, and the command of labor and people. In essence, contemporary patterns in the distribution and use of resources were instituted—the incipient political ecology of the region. In addition, through its involvement with the mining and banana industries beginning in the 1880s, the Honduran government initiated its prevailing role as ally of foreign (especially U.S.) corporations, particularly in disputes with local communities over issues of property, labor, and legal rights. In the immediate post–World War II period, other powerful international actors, including the World Bank, the IMF, and the IDB, became vital participants, by financing the necessary means (infrastructure, markets, and credit) through which the region was articulated into the nation and the world, and which fundamentally altered the ways in which international agricultural capitalism and the state could affect the natural resources and people of the south.

A limited degree of vertical and horizontal segmentation of society, at both the regional and local level, occurred as well. The southern foothill and highland areas were populated by small, medium, and larger landowners, with differential access to natural resources, who produced for themselves and for the growing market, as well as by a few landless families, mostly confined to border areas. In the lowlands, the relatively small number of extremely large landowners and the few regional elites with family histories that extended back to colonial times, who owned the most productive agricultural lands, were joined by an expanding merchant class, public and private sector service personnel, and a general class of urban workers.

With international assistance, the Honduran government functioned as an agent of development by improving access to credit, establishing markets, and constructing the requisite regional infrastructure. A number of mechanisms to regulate the emerging and more complex, regional economy, were founded, including a variety of public and private sector economic and political institutions (e.g., banks, government offices, grain merchants, and agricultural supply stores) which linked the region to the nation and the world. These measures afforded large landowners the incentive to produce for the world market. A domestic, capitalist agricultural sector emerged in the 1950s, and by the mid-1970s, large foreign agricultural enterprises began investing in the area, competing with regional capitalists for land and labor. From mid-century to the present, diversification and growth of agricultural production for export characterized the southern Honduran economy. Cotton, then sugar and livestock, were the primary commodities involved in the transformation of the south. In the mid-1970s, these products were supplemented by sesame and melons and later by a wider variety of so-called "nontraditionals," especially cultivated shrimp. For the Honduran government in persistent fiscal crisis and tied to foreign interests, struggling with the repayment of mounting external debt has taken priority over conserving natural resources. Export commodities such as cotton, cattle, melons, and shrimp, attract international assistance and investment, and increase foreign exchange earnings—whatever their social and environmental costs.

Recent Trends in the Honduran Economy

Recent government efforts to diversify and expand agriculture for export, in order to augment foreign exchange, are more comprehensible in light of Honduras' deteriorating economic conditions during the 1980s. Evidence of the international economic crisis emerged in Honduras in 1981, and intensified throughout the decade. The balance of payments and the national treasury suffered imbalances; there were significant constraints in supplying imported inputs; and private investment dropped as a result of the region's political and social problems as well as because of disturbances in the exchange and monetary systems. This situation was aggravated by the economy's vulnerability to external fluctuations affecting the demand and price of its traditional export products such as bananas and coffee (Stonich 1992).

Between 1980 and 1989, the real growth rate of the Gross Domestic Product (GDP) was 2.3 percent per year, below the population growth rate of 2.9 percent, which resulted in a decline in the per capita GDP and led to concomitant drops in wages and in living conditions (USAID 1990: 3). Rural households were disproportionately affected due, in part, to the substantial decline in agricultural prices relative to nonagricultural prices which also discouraged additional investment in agriculture. At the same time, although the real official wage rates of rural agricultural laborers rose slightly from 1974 to 1981/82, they declined during the rest of the decade, so that by the late 1980s, they were scarcely above their 1974 levels. There is considerable doubt, as well, as to whether most agricultural workers receive the official minimum wage. According to estimates made by USAID, from 1980 to 1987, the total real purchasing power of agricultural households declined by about 10 percent while in per capita terms the decline was about 27 percent (1990: 4).

While the Honduran economy registered 4 percent real growth in 1988, the economy deteriorated significantly the following year (USAID 1991b). By 1989, the Honduran external debt of US$3.3 billion was 120 percent of the annual GDP—greater, in per capita terms, than the debt of either Brazil or Argentina (CSM 1990). By late 1989, the major financial lending institutions (The World Bank, The International Monetary Fund, and the Inter-American Development Bank) placed Honduras on the list of countries that were ineligible for new loans because of overdue payments on earlier credits, as well as because of the Liberal government's reluctance to continue its economic adjustment program. Also in 1989, for what it perceived as a lack of commitment to a sound economic reform program, USAID did not release the US$70 million in Economic Support Funds (ESF) to brace Honduras' balance of payments that had been approved.[1]

Economic liberalization was an essential component of the platform of the National Party that came into power in early 1990. The new government of President Rafael Callejas, with the support of his new legislative majority, quickly initiated a program of sweeping economic reforms and signed a Framework Trade and Investment Agreement with the U.S. under the Enterprise for the Americas Initiative (EAI). In June 1990, the IMF, the World Bank, the IDB, the U.S. and Japan collaborated on a US$246 million financing package which cleared Honduras' official arrears to the international financial institutions (USAID 1992: 633, 782).[2] The major reforms of the economy were both in line with the demands of major creditors and designed to make Honduras more attractive for investors and hence promote exports: the national currency (the Lempira) was devalued by 100 percent, and a crawling peg rate of exchange was adopted; protective import tariffs were slashed from 135 percent to 20 percent; and investment regulations—both for foreigners and national entrepreneurs—were simplified (LARR 1991a, 1991b).

Fiscal deficit reduction actions included decreased public spending (including the layoffs of approximately 10,000 government workers, about 20 percent of government employees, in January 1991), the elimination of subsidies, increased water and energy tariffs, and the modification of prices to actual market values (LARR 1991a). The ensuing rise in the cost of living, further deteriorated the economic circumstances of the most vulnerable sectors of Honduran society, whose minimum wages remained unchanged and who were also most affected by the sharp rise in unemployment. The fall in real salaries compared to price increases for basic food products and growing unemployment, led to reduced income, in turn, leading to worsening health, nutrition, education, and housing indicators (CONAMA 1991: 57–59). Despite the severe effects of the structural adjustment program on the poor, the ideology of the ruling party and the ongoing economic crisis make it highly unlikely that the national government will redirect its policies away from attempting to expand export production.

The Cotton Boom

It was cotton cultivation that transformed agrarian patterns of production in southern Honduras (Stares 1972: 35; Durham 1979: 119; Boyer 1982: 91). Although cotton had been grown in the area since pre-conquest times, large scale commercial cultivation of cotton was introduced in the late 1940s and early 1950s by Salvadorans who brought seeds, chemicals, machinery, and their own labor force into the area, via the Pan American Highway, which runs through the Pacific coastal plain. Aggressive Salvadoran farmers secured Honduran bank loans, rented (or purchased) large tracts of land from Honduran owners, and began commercial production. They were joined by Honduran farmers who first began producing on a minor scale but who, by 1960, expanded production and formed their own ginning and marketing cooperative. When the Salvadorans were expelled from the country after the Soccer War in 1969, their property was confiscated and became available to the Honduran growers (Stonich 1986: 118).

A principal effect of the cotton boom was to exacerbate inequalities in access to land. Large landowners revoked peasant tenancy or sharecropping rights, raised rental rates exorbitantly, and evicted peasants forcibly from national land or from land of undetermined tenure (K. Parsons 1975; Durham 1979; Boyer 1982). Thus one of the effects of increased cotton cultivation was to displace poor farmers from the more suitable agricultural lands in the south. The cotton boom, however, provided a substantial number of seasonal jobs during the two month harvest season (approximately 76.7 workdays per hectare per year) because the long staple cotton grown in the region was generally picked by hand.

In response to the boom and bust cycles of the international cotton market, the amount of land planted in cotton, cotton production, and the number of jobs in the industry, fluctuated considerably between the 1950s and the 1980s (Stonich 1992). Mounting cotton prices during the 1960s sustained the escalating costs of production, brought about by spiraling pesticide and fertilizer use. By 1966, cotton production reached a peak of 33,000 metric tons nationally, the area under cultivation climbed to 18,200 hectares, and the number of jobs reached 20,000 (Boyer 1982: 91; FAO-PY, various years). Declining world prices, coupled with severe infestations of pesticide resistant insects, abruptly ended the cotton boom and thousands of farmers were forced into bankruptcy or into the cultivation of other crops, while thousands of harvesters were obliged to look for other employment. By 1970, cotton production fell to 9,000 metric tons and the land area planted in cotton declined to 4,000 hectares. Improved market conditions in the late 1970s stimulated cotton cultivation once again, and the area in cultivation climbed to 13,000 hectares by 1980 before beginning a steady decline to 1,600 hectares, nationally, in 1990 (Cotton Cooperative Annual Report 1978–1982; ECLAC 1987; FAO-PY, various years). By 1990, cotton had essentially disappeared from the southern lowland plains, with only about 50 hectares projected for the 1991 season (Murray 1991).

As in El Salvador and Nicaragua, commercial cultivation in Honduras involved the combination of manual labor (in harvesting) and mechanization (in land preparation, planting, cultivation, and aerial spraying). Cotton cultivation along the Pacific coastal plain is dependent on the heavy use of chemical inputs (especially insecticides and fertilizers) which have significant environmental repercussions especially when these inputs are inappropriately used. The indiscriminate use of pesticides in what previously were cotton growing regions, and now are centers of nontraditional crop production or pasture, remains one of the most pervasive environmental contamination and human health problems throughout Central America (ICAITI 1977; Weir and Shapiro 1981; Bull 1982; Bottrell 1983; Boardman 1986; Leonard 1987). The environmental costs of cotton production along the Pacific coastal plain of Central America—including southern Honduras—have been extreme: fertile soils have been depleted through continuous cultivation; soil erosion has been accelerated due to the general lack of planting leguminous ground cover capable of absorbing rainfall from the heavy thunderstorms that occur during the rainy season; the growth of harmful insects has been encouraged because of previous strategies aimed at the total eradication of cotton pests with unregulated applications of DDT and other chlorinated insecticides; and land and water contamination, as well as high levels of pesticide residues in food supplies, threaten human health (Williams 1986; Leonard 1987; SECPLAN/USAID 1989; Murray 1991).

The Cattle Boom

Opportunities for the production of other export commodities expanded concurrently with the decline of cotton. Although the promotion and growth of the export commodities that succeeded cotton were not limited to livestock, it was the expansion of the cattle industry that had the most extensive and devastating environmental and social consequences. Before the mining and banana booms of the 1880s, cattle were Honduras' major export (Stonich 1986: 112). Production stagnated or declined during the years of the initial banana boom, the world-wide depression, and World War II, then increased dramatically in the post-war period. International and national development efforts after the war fostered the growth of the livestock industry. These ventures accelerated during the 1960s as part of the Alliance for Progress which helped to promote the livestock boom throughout Central America. Export quotas to the United States were increased, promotional efforts were undertaken to stimulate the beef trade and to modernize beef production, and credit programs were established to help expand the production of export commodities such as beef. Between 1960 and 1983, 57 percent of the total loan funds allocated by the World Bank for agriculture and rural development in Central America supported the production of beef for export (computed from Table 4.1 in Jarvis 1986: 124). During that period, Honduras obtained 51 percent of the total World Bank funds that were disbursed in Central America—of which 34 per-

Recently cut *ceiba* tree surrounded by grazing cattle in the Choluteca lowlands.

cent were for livestock projects (computed from Table 4.1 in Jarvis 1986: 124). Between 1947 and 1988, the number of head of cattle rose 220 percent in Honduras and 152 percent in Central America (computed from FAO-PY, various years). In a context of declining agricultural commodity prices, high labor costs, unreliable rainfall, and international and national support for livestock, landowners reallocated their land from cotton or grain cultivation to pasture for cattle. In Honduras, land reform programs also encouraged investment in livestock. Landowners who feared expropriation of unutilized fallow and forest land, fenced it, planted pasture, and stocked it with cattle as a way of establishing use without increasing labor inputs (Jarvis 1986: 157).

The growth in livestock production that occurred in southern Honduras between the 1960s and the early 1980s took place disproportionately on larger farms: between 1952 and 1974, the percentage of cattle owned by farms of 100 or more hectares increased from 31 percent to 56 percent of all cattle owned in the south (Stonich 1986: 114–115). Expansion took place not only in the lowlands and foothills, where cattle raising traditionally occurred, but also in the highlands where many of the wealthier peasant farmers augmented cattle production. Because of the land-extensive system of cattle raising in the south (1 hectare per head) and the low labor demands (6.3 workdays per hectare per year), the spread of cattle raising could not absorb the growing landless and land-poor population and, in fact, intensified their expulsion from national and private lands (White 1977: 126–156; Stonich 1986: 139–143; Boyer 1987; Howard-Ballard 1987).

In general, paralleling trends that took place throughout Central America, beef exports from Honduras reached their peak in the early 1980s. Between 1981 and 1988, beef exports declined by 50 percent and beef production dropped 30 percent, in response to reduced demand, falling prices, reduction in Honduras' export quota to the United States in 1984, and import restrictions reflecting tighter quality control regulations for meat imports introduced by the United States in 1983 (LACR 1986; USDA 1987; Stonich 1992). These national trends were apparent in the south, where by 1981, Honduran producers reintroduced on-the-hoof trade with El Salvador and Guatemala. By 1982, meat packing plants frequently were shut down because of the lack of an adequate volume of cattle to operate efficiently (Stonich 1986: 117–118).

The environmental corollary of the cattle boom was the reallocation of land from forest, fallow, or food crops, to pasture. The transformation of forest to pasture that took place throughout Central America as part of the livestock boom is well known (see J. Parsons 1976; Myers 1981; Nations and Komer 1983; DeWalt 1985; Shane 1986; Williams 1986). However, it is essential that this profound ecological transformation be understood as part of a more comprehensive set of processes: i.e., each shift in the promotion and the relative importance of export commodities has been accompanied by significant ecological consequences. Nonetheless, the extent of the interrelated mechanisms of forest destruction and pasture expansion was exceptional. Although 50 percent of Central America was in forest in the early 1960s, forest land had declined to 41 percent by 1972, to 34 percent by 1982 and to 30 percent by 1987 (Figure 4.1). Between the early 1960s and the late 1980s, the percentage of forested land decreased by 51 percent in Honduras, 55 percent in El Salvador, 43 percent in Costa Rica, 42 percent in Nicaragua, and 37 percent in Guatemala (computed from FAO-PY, various years). Simultaneously, the percentage of land in pasture escalated throughout Central America—from 17 percent in the early 1960s to 30 percent in 1987. Southern Honduras was part of this broader trend (Figure 4.2). Between the early 1950s and the mid-1970s pine and deciduous forests decreased by 44 percent and fallow land (vital to regenerative shifting cultivation systems) plunged by 58 percent, while pasture land rose by 53 percent (Stonich 1989). By 1974 deciduous and pine forests covered only 13 percent of the region while pasture comprised over 60 percent of the total land area (Stonich 1989).

Changes in Land Tenure
and in the Distribution of Land

The expansion of commercial agricultural production on large landholdings had far reaching and long-term effects on the availability of farmland for small farmers. Until mid-century, smallholder farmers had access to land in a variety of tenure arrangements. In 1952, smaller farms of less than 10 hectares accounted for 73.5 percent of all farms in the region, but a disproportionate percentage of farms in different tenure categories: 59 per cent of privately owned farms, 88.6 per cent of all rented farms, 98 percent of all sharecropped farms, 93.7 percent of all farms on national land, and 80 percent of all farms located

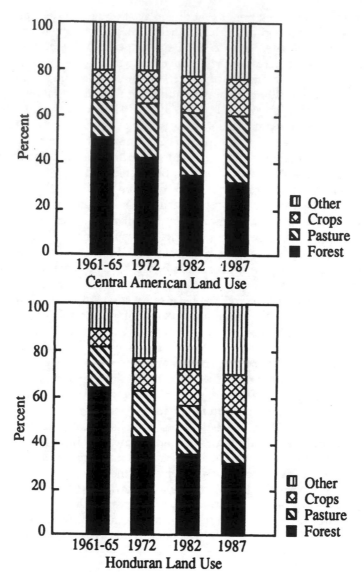

FIGURE 4.1 Changes in Land Use in Central America and Honduras: 1961–87. Source: FAO-PY, various years.

on community managed, *ejidal* land (Stonich 1986: 140). On the other hand, while only 28 percent of farms of under ten hectares were privately owned, 30 percent were either sharecropped or rented, and an additional 26 percent were on public (national) or *ejidal* land (Stonich 1986: 140).

The presence of small peasant farmers and their families on rented, national, and *ejidal* land, that could potentially be put into commercial production, impeded large scale capitalist expansion, and beginning in the 1950s

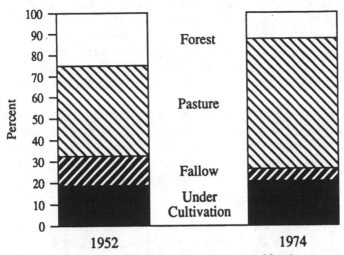

FIGURE 4.2 Changes in Land Use in Southern Honduras: 1952 and 1974. Source: DGECH 1954, 1976.

TABLE 4.1 Distribution of Farmland by Farm Size Category in Southern Honduras: 1952 and 1974

	Percentage of Total Farms				Percentage of Total Farmland Area			
	1952		1974		1952		1974	
Size[a]	%	Cum. %	%	Cum. %	%	Cum. %	%	Cum. %
<1	12.0	12.0	21.0	21.0	0.6	0.6	1.0	1.0
≥1, <5	48.0	60.0	47.0	68.0	8.8	9.4	9.0	10.0
≥5, <10	16.9	76.9	13.0	81.0	8.6	18.0	8.0	18.0
≥10, <20	10.6	87.5	8.8	89.8	10.4	28.4	10.4	28.4
≥20, <50	7.7	95.2	6.0	95.8	17.0	45.4	14.8	43.2
≥50, <200	3.2	98.4	2.5	98.3	20.0	65.4	20.0	63.2
≥200, <500	0.6	99.0	0.5	98.8	11.0	76.4	15.0	78.2
≥500	0.4	100.0	0.2	100.0	23.2	100.0	22.0	100.0

[a] Farm size categories in hectares.

Source: DGECH 1954, 1976.

large landowners used a variety of methods to remove them (White 1972: 126–156; 1977: 173). Along with capitalist expansion, regional land values increased considerably, affecting not only the purchase price but rental costs as well. The relatively wide variety of rental arrangements, that often included payment in kind, were replaced with cash-only agreements. Rental fees jumped from US$1.50–2.50 per *manzana*[3] to between US$10–20 per *manzana*, far above the amount most small farmers were able to pay (K. Parsons 1976: 15). In the 1950s and early 1960s, peasants often were forcibly evicted. White (1972, 1977) and Parsons (1976) concluded that much of the expansion in cotton and livestock production occurred through the illegal expansion of large haciendas often into national and ejidal land that was already occupied by peasant families. Earlier, many of the families had been granted titles to the land, and when they resisted eviction, they were forcibly removed often with the assistance of local authorities and the police (White 1972: 130–156). Between 1952 and 1965, the total ejidal and national land declined from 22 percent to 15 percent of available farmland, then dropped to 10 percent by 1974, while the percentage of privately owned land increased concomitantly during the same period (computed from DGECH 1954, 1968, 1976).

The forced evictions, exorbitant rental fees, and declines in available national and *ejidal* land contributed to decreases in the amount of land to which small farmers had access. These factors, along with the competitive allocation of land from food crop production to export crop production and the significant increases in regional population, concentrated landholdings and excluded small holders more than ever before. Table 4.1 compares the distribution of the number of farms and the area of farmland, to the percentage of farms in various size categories, in 1952 and in 1974. It shows a high concentration in the distribution of farmland at both time periods. In 1952, large farms of more than 200 hectares represented only 1 percent of all farms and held 34 percent of all farmland, while small farms of less than five hectares (60 percent of all farms) controlled less than 10 percent of all farmland. By 1974, large farms of more than 200 hectares comprised 0.7 percent of all farms and controlled 37 percent of all farmland. At the same time the number of farms of less than 5 hectares increased to 68 percent of all farms and controlled approximately the same amount of land (10 percent) that farms in that size category controlled in 1952. The greatest change in the distribution of farms between 1952 and 1974, was in the percentage of farms of less than one hectare; from 12 percent of all farms to 21 percent. During the same period, the amount of farmland held by farms of that size increased only .4 percent. Together these data suggest a trend toward greater concentration of land holdings, particularly in the highest and lowest categories. The actual concentration of landholdings is even greater than shown in Table 4.1 because these data do not take into account the fact that an estimated 30 percent to 34 percent of rural families were landless by 1974 (USAID 1978: 236).

Peasants in the south were not compliant to the powerful forces affecting the region and have a history of resistance (White 1977; Boyer 1982). Bolstered by French Canadian priests and nuns who were part of the Catholic Social Movement, the efforts of southern peasants helped prompt an agrarian reform movement that by 1975 had redistributed approximately 18,000 hectares of land to 3,500 families in the south (Boyer and Church 1987).[4] During the same period, however, large landholders succeeded in acquiring 13,000 hectares of land, in effect negating two-thirds of the gains made through the agrarian reform (Boyer and Church 1987). By the early 1970s, the National Congress, influenced by large cattle ranchers, threatened to expel the French Canadians whose activities had become increasingly radicalized around issues of agrarian reform (White 1977). Since then, land distribution in the region has declined to almost nothing. According to the regional director of INA in 1985, the majority of land on reform communities was characterized by poor soils, steep slopes, and extreme erosion, and many reform communities were renting their land to cattle ranchers who used it for pasture.[5]

Agrarian Change and Food Security

For a time in the post-war period, there were increases in the production and export of basic food crops (corn, sorghum, and beans). By the early 1960s, under the patterns of trade that emerged as part of the newly established Central American Common Market (CACM), Honduras exported considerable quantities of corn and beans to other Central American countries, especially to El Salvador. The production of basic grains took place largely on small farms and did not expand at a sufficient rate to keep pace with foreign and increased domestic demand brought about by population growth and by the growth of the urban centers of Tegucigalpa and San Pedro Sula. This resulted in a decreased grain supply in the rural areas by the early 1960s (O'Brien Fonck 1972: 92). Except for grains, most of the market expansion associated with export growth supported large farmers. In the early 1950s, with the exception of coffee, export crops were primarily produced on large farms while the majority of traditional basic grains, the major human food crops, were produced on small farms of less than 20 hectares (OAS 1962). By 1974 (the most recent year for which data are available) such farms comprised 88 percent of all farms and continued to produce more than 70 percent of basic food crops (Stonich 1991b). Since 1950, there has been a marked stagnation in the production of domestic food crops nationally while the production of export crops increased (Stonich 1991). In the south, as shown in Table 4.2 and Table 4.3, the area cultivated and the total production, per capita, of corn, sorghum, and beans fell between 1952 and 1989. By 1989, the area cultivated in corn declined to 49 percent of its 1952 level, while per capita production fell to 28 percent. The drop in the area in cultivation and per capita production of beans was more steep: to 15 percent of the cultivated area and to 5 percent of per capita production. The declines in the cultivation and production of sorghum were less severe because sorghum is also grown as livestock feed. Table 4.2 also provides examples of the expansion and contraction in the export sector, especially in the boom and bust cycles of the cotton industry and the current stage of expansion in melon production.

The Honduran government contributed to the stagnation of basic grains by placing price ceilings on basic staples as a way of maintaining cheap food for the growing urban population. Between 1970 and 1981, real farm prices for corn, beans, rice, sorghum, eggs, milk, chicken, pork, and a variety of other crops declined in spite of stagnating production (Larson 1982). Between 1980 and 1986, real farm prices for corn, sorghum, and beans remained the same or declined (IHMA 1987).

On the other hand, the government continued to provide subsidies to encourage export agriculture for foreign exchange. Between 1970 and 1980, 77 percent of agricultural loans disbursed through the national banking system were allocated to export crops and livestock (52 percent and 25 percent respectively), while credit for basic grains comprised only 13 percent, and other crops for domestic consumption, 10 percent (Ponce 1986: 146–150). If a distinction is made between commercial banks and state banks the disparity is even greater: commercial banks allocated about 81 percent of their agricultural loan funds for export crops and livestock during the same period (Ponce 1986: 146–150). Although public development banks held 82 percent of loan funds allocated for basic grains, they, nevertheless, preferred lending for export commodities and to producers who offered better loan guarantees (Ponce 1986: 146–150). Between 1980 and 1986, the major proportion of Central Bank (BANTRAL) funds were apportioned to the industrial, commercial, and service sectors (more than 70 percent) while the agricultural sector was allocated approximately 21 percent of funds (BANTRAL 1987). Of the agricultural loans disseminated, only 2 percent were allocated to basic grains (corn, beans, sorghum, rice), while the rest was allotted to export commodities (BANTRAL 1987). A similar pattern of capital disbursement is evident in the loans provided by the National Agricultural Development Bank (BANADESA): between 1980 and 1982, only 5 percent of the loan funds went to the production of corn, beans, and sorghum with the rest going to livestock, cotton, and sugar—commodities produced by large farmers primarily for export (Stonich 1986: 124–127). The distribution of funds changed somewhat in the early 1980s reflecting government policies aimed at making Honduras more self sufficient in the production of basic food grains. In 1986, the percentage of agricultural loans apportioned to basic food crops was 25 percent of all agricultural loans, with most of the remaining funds allocated to livestock, cotton, sugar, and various fruits and vegetables produced mainly for foreign markets (BANADESA 1987). However, during the 1982 to 1986 period, the value of loan funds allotted to basic grains actually declined 12 percent while livestock loans increased 216 percent (Stonich and DeWalt 1989).

Despite the critical need for improved access to institutional credit for smallholder farmers, a 1989 survey of Honduran grain producers conducted by an agency of the European Economic Commission—the Committee to Support the Economic and Social Development of Central America—neither BANADESA nor the commercial banks met this demand—less than 10 percent of basic grain farmers received institutional credit (USAID 1990: 13). A similar study, done somewhat earlier, found that less than 10 percent of all small farmers obtain institutional credit (USAID 1988). More recent IMF inspired government policies, including raising the discount rate to 25 percent and, hence, the

TABLE 4.2 Area Under Cultivation in Major Crops in Southern Honduras: 1952, 1965, 1974, 1982, and 1988/89

Crop	1952[a] Area[b]	%	1965 Area	%	1974 Area	%	1982 Area	%	1988/89 Area	%
Corn	37,569	100	34,128	91	48,815	130	38,089	101	18,533	49
Sorghum	20,197	100	23,641	117	30,830	153	23,224	115	18,106	90
Beans	4,111	100	4,154	101	5,250	128	3,496	85	596	15
Cotton	883	100	13,016	1,474	3,249	368	3,124	354	1,180	134
Sugar	—	—	—	—	5,896	100	1,950	203	—	—
Melon	—	—	214	100	296	138	1,084	507	4,500	2,103
Sesame	475	100	241	51	3,249	684	429	90	1,808	380
Rice	1,046	100	—	—	1,392	133	3,371	322	417	40

[a] Index 1952 = 100.
[b] in hectares.

Source: DGECH 1954, 1968, 1976; SRN 1982; SECPLAN 1990.

TABLE 4.3 Per Capita Production of Basic Food Crops in Southern Honduras:
1952, 1965, 1974, 1982, 1989/90

Crop	1952[a]		1965		1974		1982		1989/90	
	N[b]	%	N	%	N	%	N	%	N	%
Corn	3.10	100	1.90	61	1.80	58	0.90	29	0.88	28
Sorghum	2.20	100	1.30	59	1.80	55	0.60	27	0.79	36
Beans	0.21	100	0.18	86	0.08	38	0.04	19	0.01	5

[a] Index 1950 = 100.
[b] N = per capita production in 100 pounds.

Source: DGECH 1954, 1968, 1976; SRN 1982; SECPLAN 1990.

actual rates at which farmers can borrow to at least 30 percent, have diminished the borrowing capacity of small farmers further—as well as garnering protest from small farmer groups (LAEB 1991).

The combination of price ceilings and limited credit contributed to the stagnation of small farm food production, the major supplier for the national market. Nationally, per capita production of basic grains declined 31 percent between 1950 and 1985 (USDA 1985). Per capita food production indexed at 100 in 1970 fell to 69 by 1989 (World Bank 1990). At the same time, the production (in thousands of metric tons) of the major export crops (bananas, coffee, sugar, and cotton) increased 172 percent (computed from data contained in ECLAC 1987 and FAO-PY 1986) and the production of beef (principally for export) rose 267 percent (computed from data in DGECH 1954, USDA Foreign Agriculture Circulars 1959–1985, and FAO-TY 1982–1986). One result, was that Honduras became a net importer of such basic staples as corn, beans, sorghum, and rice (FAO 1984: 103–4; Stonich 1986: 124–126). Between 1974 and 1986, cereal imports climbed from 52 metric tons (19.3 kilograms per capita) to 122 metric tons (27.1 kilograms per capita); food aid in cereals expanded from 31 metric tons (11.5 kilograms per capita) to 135 metric tons (30 kilograms per capita); and the percentage of food aid in cereals to imports of cereals, escalated from 60 percent to 111 percent (Stonich 1991b). Although, between 1981 and 1986, the amount of corn imported into the country declined 17 percent (from 18,100 tons to 15,100 tons), this decrease was exceeded by the 73 percent boost in the amount of imported wheat flour and processed wheat products (from 75,000 tons in 1981 to 130,000 tons in 1986) (ADAI 1987). At the level of the consumer, there has been a decline in the price of wheat flour relative to the retail prices of most other basic foods, and the real retail price of wheat flour declined by 41 percent between 1975 and 1985 (Garcia et al. 1988: 130–131). This trend in prices (brought about, in part, by government pricing policy on wheat imports) apparently influenced a change in average consumer diets toward more use of wheat products which are substituted for corn and other traditional staples (Garcia et al. 1988: 131). In 1990, Honduras resorted to a record volume of food imports and this growing dependence on imported wheat should be viewed in the context of increasing dependence on imported food (UNICEF 1991).

The Effects on Diet and Nutrition

Current average nutrition levels in Honduras are slightly below their 1980 levels, and evidence suggests that despite thirty years of economic growth, a majority of Hondurans find themselves less able now than in the mid-1960s to meet their minimum nutritional needs (USAID 1990: 3) Trends in food production and purchasing power have certainly worsened widespread calorie deficiency, especially in rural areas where the deficit is estimated to be approximately 20 percent (Garcia et al. 1988). Overall, from the 1960s to 1986, the availability of domestic staples, calculated in terms of daily per capita supply, declined by 31 percent (computed from World Bank, various years). According to national food balance sheets, between the mid-1960s and the mid-1970s daily per capita available food energy (i.e, calories) declined 25 percent; to 90 percent of the daily average per capita requirement (Valverde 1986). By 1969 according to a survey by the Institute of Nutrition for Central America and Panama (INCAP), rural Honduran families were meeting, on an average, only 89 percent of daily energy requirements (1969). Food consumption surveys carried out by the national planning agency (SAPLAN) indicated that per capita daily energy intakes fell to only 80 percent of daily needs between 1972 and 1979 (1981). CEPAL (1982) estimated that by the early 1980s, 68 percent of Honduran families did not have the purchasing power to satisfy their food requirements. While the incidence of malnutrition according to the National Nutritional Survey of 1987 was 38 percent of children under five, other studies suggest that in some areas of the country 70 percent of children under five suffer some degree of malnutrition (USAID 1990: 3). Conditions are worse in rural areas because the majority of poor households are rural—the average urban household is estimated to have four times the per capita income of the average rural household, and consequently tends to be better nourished (Garcia et al. 1988). The economic decline that distinguished the Central American economies recently has deepened hunger among the poor. Food security is beyond their means as per capita stocks of the staples consumed by the poor dwindle and as their ability to buy food also decreases (Stonich 1991b).

The transformation of the agrarian structure in the south—including profound conversions in land use, the drastic fall in per capita production of basic grains, growing land concentration and landlessness, as well as inadequate employment opportunities to earn cash to buy food—intensified undernutrition. The nutritional consequences of agrarian transformation on people in the south were evident quickly, and assessed by several nutritional studies beginning in the mid-1960s. On the basis of their 1966 survey of Honduras, which included the south, INCAP concluded that 76 percent of preschool children in the region suffered from protein-energy malnutrition (INCAP 1969). Based on his study of the rural south, Stares concluded, in 1972, that the INCAP study underestimated the degree of undernutrition because of its focus on municipal centers rather than on more remote rural areas (1972: 25–28). In his ethno-

graphic study conducted in the late 1970s, Boyer found that the situation had
deteriorated, and estimated that, by that time, half the highland peasants were
deficient in calorie consumption (Boyer 1982: 246). His conclusion was
strengthened by a study done, in 1980, by SAPLAN which compared studies
of average diets between 1966 and 1979 and concluded that a decrease in the
overall availability of energy (calories), iron, and vitamins A and C had ensued during the period (SAPLAN 1981). Further, in their 1978 nutritional survey of three regions of Honduras, including the south, SAPLAN concluded
that 41 percent of all southern families did not have the purchasing power to
meet minimum subsistence needs and for much of the year were hungry, and
moreover, families living in "semi-urban" (i.e., marginal or squatter) communities consumed even fewer calories than did rural families (SAPLAN 1981).
As discussed in Chapter 2, a regional breakdown of the results of the National
Nutritional Survey conducted in 1986 revealed that 35 percent of all first graders in the south were stunted (SAEH/INCAP 1987). Local level studies in the
highland communities, that are the focus of the following chapters, showed
that 65 percent of children under 60 months of age were stunted, 14 percent
were wasted, and more than 50 percent of all households in some communities
did not meet their calorie requirements (DeWalt and DeWalt 1987: 39).

The New Nontraditionals:
Shrimp and Melons

With the decline in other economic sectors, the Honduran government began
initiating policies to enhance nontraditional export growth in the early 1980s.
It declared 1987 as "the year of the exports" and undertook several measures
to stimulate private investment and export production—especially "nontraditionals": import taxes for inputs used in export products were eliminated; exporters were allowed to keep part of their export earnings for direct purchases;
and investment policy and export regulations were simplified (USAID 1989a).
The World Bank, USAID, and the EC, among others, encouraged the shift to
nontraditionals through the infusion of new projects, loans, and funding
(Heffernan 1988). With the financial support of USAID, a number of non-profit
organizations designed to promote Honduran exports through the creation of
information and business networks were established: the Honduran Federation of Agricultural and Agro-Industrial Producers and Exporters
(FEPROEXAAH/FPX), the Foundation for Entrepreneurial Research and Development (FIDE), and the Honduran Agricultural Research Foundation
(FHIA).

Between 1980 and 1987, the value of nontraditional agricultural exports (including agricultural crops, agro-industrial products, and shrimp) grew from
US$65.7 million to US$107.8 million (USAID 1989c). During that time period,
the value of nontraditional agricultural crops rose from approximately US$21
million to US$32 million and the value of exports of shrimp and lobster more
than doubled (from US$23.4 million to US$50 million). By 1987, income from
the export of shrimp ranked third after bananas and coffee in total export earnings for Honduras supplanting the position that beef exports had held previ-

ously (Stonich 1991c). Nontraditional export development in the south has emphasized irrigated melon production in lowland areas and shrimp maricul-ture in coastal zones along the Gulf of Fonseca (Stonich 1991c). Continuing the pattern established in the post-war period, the Honduran government, in col-laboration with the U.S. and other foreign interests, currently is attempting to stimulate shrimp and melon exports through improvements in infrastructure including the construction of a regional airport for the city of Choluteca, repair and improvements of the Pan American Highway, and the construction of two industrial parks in the cities of Choluteca and San Lorenzo (*Honduras This Week* 1992: 1, 19).

The Shrimp Boom

The decline in the importance of beef as an export commodity occurred con-currently with the commencement of cultivated shrimp as the most important new export product—part of the Central American regional attempt to en-courage the exploitation of marine resources as part of the diversification of exports. Principal investors have included transnational corporations, gov-ernment and military leaders, as well as consortiums of private investors. As in the rest of Central America, the enlargement of shrimp farms was financed by national and international, private and public capital (USAID 1985, 1989c; SECPLAN/USAID 1989). This included direct financing through loans (e.g., between 1986 and 1989 USAID provided US$7 million for seven large farms and the quasi-governmental Rural Technologies Program channelled an addi-tional US$1 million to medium-scale producers) and indirect funding in the form of incentives to foreign investors (e.g., through certificates of export pro-motion and foreign exchange and bonds for tax payments) (USAID/FEPROEXAAH 1989). The total production of cultivated shrimp grew from 130 metric tons to 2,225 metric tons (1611 percent) between 1978 and 1988 and the area in production grew from 1,450 hectares to 5,500 hectares (280 percent) in the three year period from 1986 to 1989 (USAID/FEPROEXAAH 1989). By 1988 production and exports from shrimp farms exceeded that from industrial fisheries (SECPLAN/USAID 1989). According to estimates by USAID, the area in shrimp farms may expand to more than 15,500 hectares by 1995 with an estimated export value of US$70–100 million (USAID/FEPROEXAAH 1989).

The environmental analysis of the USAID funded, Investment and Export Development Project, asserts that the, "pitifully little research on the natural resources of the Gulf of Fonseca's estuaries, mangrove forests, and mudflats" makes it impossible to evaluate adequately the significance of environmental changes emanating from the ongoing expansion of shrimp farms (Castañeda and Matamoros 1990). Without such information it is impossible to predict precisely the consequences of further loss of mangroves, the increased sedi-ment load in the water column, the construction of roads through the estuaries on levies instead of bridges, and the altered oxidation reduction potential from the loss of the 14,000 hectares of mudflats that have already been con-signed and are scheduled for conversion to shrimp farms by 1995 (Castañeda and Matamoros 1990).[6]

Small-scale, solar salt-evaporation and shrimp-farm ponds in Choluteca. The ponds are used to make salt during the dry season and to cultivate shrimp during the rainy season.

The Distribution of Costs and Benefits at the Local Level. The unequal distribution of coastal resources associated with concessions granted by the Honduran Department of Tourism (SECTUR) is evident: of the 57 concessions granted by 1988, 25 (44 percent) were small holdings (1–70 hectares) with access to only 723 hectares of land (2.6 percent of total concessions) while the eight largest concessionaires had rights to 19,535 hectares (69 percent) (SECPLAN/USAID 1989). According to a study done by economists at the Honduran National Autonomous University (UNAH), by 1991, five farms owned or had concessions of approximately 1,000 hectares or more: Granjas Marinas (5,055 hectares); Aquamarina Chismuyu (3,000 hectares); Aquacultivos de Honduras (1,540 hectares); Aquacultura Fonseca (957 hectares); and Cumar (934 hectares). Of these, only Aquamarina Chismuyu had a concession on what had been private land; the rest were national lands. This same study estimated that approximately 72 percent of the coastal land utilized for shrimp farms is "national" land (Banegas Archaga et al. 1991).

The unequal distribution of shrimp farms in operation is clear as well. By 1990, approximately 76 farms operated in the south. Of these, five were large farms of more than 250 hectares each (total of 4,200 hectares), 21 were medium farms ranging from 20 to 250 hectares in size (total of 1,550 hectares), and 50 were small family farms of less than 20 hectares (total of 250 hectares). The five largest farms (6 percent of the total farms) controlled 70 percent of the land in production, the 21 medium farms (28 percent of all farms) managed 26 percent of the farmland, and the 50 smallest farms (66 percent of all farms) had access to only 4 percent of the land in production (Mejia 1991). The rationing of concessions and permits to build ponds, as well as the large capital outlays required for intensive (and semi-intensive) operations, promote investment by government and military officials and by urban elites, and limit investment by less affluent and influential segments of Honduran society. The majority of farms under 70 hectares have been organized by urban investors with no previous experience in the industry and with insufficient capital to establish production. Many lack aeration equipment, technical assistance, and access to transportation and marketing infrastructure. Such constraints are likely to result in the further concentration of shrimp farm holdings, as these marginal operations are unable to maintain production and repay loans (Casteñeda and Matamoros 1990).

   The Effects on Coastal People and the Environment. Limited information from communities in the region which are most affected by the development of shrimp mariculture suggests that the household economies in these communities are organized much like those in more agriculturally oriented communities in Central America: i.e., they are remarkably flexible, dependent on remittances, and can shift among resources in response to changing market conditions and local resource availability (Stonich 1991c). Like these communities, there appears to be considerable variability in socioeconomic differentiation as well; while approximately 25 percent of households own the essential fishing gear (e.g., boats, nets, and motors), the remaining 75 percent of households work as hired laborers for their more affluent neighbors. In this regard, ownership and control of the means of production (land, agricultural implements, boats, motors, nets, etc.) are similar to that found among agriculturalists and are centered in the household. Although shared labor does occur, such sharing generally takes place among members of extended families. At the same time the "tradition" of joint community action to acquire resources also exists. For example, twenty one families from one coastal community currently are engaged in a land occupation on *Isla La Carretal*. By August 1992 they had occupied and planted with corn, beans, watermelon, squash, melons, and manioc, 40 *manzanas* from a total of 300 *manzanas* that they plan to clear and plant.

   Such groups are responding to the reduction of resources available to them which jeopardize their livelihoods. Until the mid-1980s, when the construction of shrimp farms accelerated, the south's mangrove ecosystems provided a source of communal resources for families inhabiting the coastal zone. Many of these families were among those that had been displaced by the earlier expansion of cotton, sugar, and livestock in the region (Stonich 1991c). Prior to the expansion of shrimp farms, the commercial and subsistence activities of these families included salt production, fishing, hunting, shellfish collecting, tannin production, and collection of fuelwood. New land tenure relationships dictated by land pricing structures have tended to benefit the shrimp farms, while simultaneously placing traditional community *ejidal* lands at risk, and fomenting land speculation. Past events in which small farmers were removed from relatively good agricultural land, often by force and with the compliance of local authorities, have been repeated on the intertidal lands which have not been cleared. Wetlands, once open to public use for fishing, shellfish collecting, and the cutting of firewood and tanbark are now being converted to private use.

Another source of friction has been the overlapping responsibilities of the various government agencies involved with different aspects of shrimp farming. Since the passage of *Decreto Ley* 968 in 1980, the Secretary of Culture and Tourism (SECTUR) has had jurisdiction over granting concessions because of its mandate to oversee state lands which boarder beaches and other tourist areas under that law. Before passage of that law, the National Agrarian Institute (INA) managed coastal lands. At the same time, the Honduran Corporation for Forestry Development (COHDEFOR) had (until recently) the responsibility for protection and rational use of Honduras' forest resources and shared responsibility for the protection of the mangroves with the Department of Renewable Natural Resources (RENARE) which also has the right to supervise fishing and aquaculture within the country. The lack of unclouded demarcations of agency responsibilities has led to dissension and confrontation. Concessions are often granted without taking into consideration environmental suitability, the competence of the applicant, or even whether the current request overlaps with previous concessions (USAID/FEPROEXAAH 1988). Conflicts have arisen among the large foreign owned operations, local medium scale entrepreneurs, and peasant cooperatives over access, and between shrimp farmers and artisanal fishers.

At the same time, the extent to which the expanding shrimp industry can increase income for significant numbers of local people is problematic. According to a study conducted in 1987 of the labor force employed by twenty-nine shrimp farms, these operations created a total of 1,130 jobs (of which only 31 percent were permanent jobs) or .785 jobs per hectare (less than one job per hectare). Of the total number of jobs, 93.5 percent were unskilled (manual laborers, assembly line workers, and security guards), 4.6 percent were administrative (administrative assistants and secretaries), and 1.9 percent were skilled (biologists and technical supervisors) (Gonzalez 1987). In response to mounting public criticism of the industry, the National Association of Shrimp Farmers of Honduras (ANDAH), whose members tend to be owners and operators of larger farms, issued their own estimate of the number of jobs created: 1.5 jobs per hectare (ANDAH 1990).[7] Whichever estimate is used, the number of jobs created is insufficient in a region where close to 100,000 households are landless or landpoor and in which the estimated unemployment rate is more than 60 percent (ADAI 1987). In addition, while the majority of jobs are unskilled in all farm size categories, the overwhelming bulk of salaries is paid to skilled employees (Stonich 1991c). The skilled positions are highly unlikely to be filled by local people. Many of the highest paying jobs in these categories are held by non-Hondurans while lower ranking administrative positions, such as secretaries, are filled by people from outside the local area (Stonich 1991c).

Diminished resources are also apparent in the declines in the number of fin-fish and shellfish captured since 1987, reported by artisanal fishers. Destruction of habitats, blocking of estuaries, and rechanneling of rivers associated with the expansion of shrimp farms encourage ecological imbalances and destruction of other flora and fauna. The purported use of *Rotenone* by shrimp farmers to eliminate other species in ponds is also directly related to loss of stocks. There are several additional interacting factors that are influencing this decline including increased sedimentation from erosion at higher elevations; a decade of drought; *El Niño* conditions; and the presence of pollutants from growing human populations and from the uncontrolled use of pesticides in the production of other export commodities (especially melons). Recent samples taken from Choluteca's shrimp farms showed relatively high levels of DDT in the young pond shrimp, which dropped below the maximum acceptable residue limits before the shrimp matured and were harvested. DDT or other pesticide residues reportedly have not yet exceeded international limits but with each new pest outbreak in melons and each incremental increase in the volume and variety of pesticides applied, the risk that chemical contamination will affect the shrimp export industry grows (Murray 1991).

Grassroots Environmental Activism. The expansion of shrimp farms has taken place over the protests of individuals and environmental groups. The most significant of these is the Committee for the Defense and Development of the Flora and Fauna of the Gulf of Fonseca (CODDEFFAGOLF). Established in 1988, principally by artisanal fishers to draw attention to the social and ecological problems associated with the expansion of the shrimp industry, the goals of the organization are to promote a balance among conservation, sustainable development, and social justice. CODDEFFAGOLF members have organized a sequence of protests: members have sent letters and proclamations to the President of Honduras, to the President of the Honduran Congress, and to the Commander in Chief of the Armed Forces; they have marched on the Honduran Congress with their demands; and they have physically blocked heavy earth-moving equipment. As an alternative to current destructive development practices in coastal areas, the group has proposed that the Honduran Congress create some variation of a national park or resource extraction reserve. The organization has generated a good deal of publicity and outside support for their goals among the Honduran public, the press, and international environmental groups.[8] Their growing renown was recognized at UNCED in 1992 where they received a Global 500 award, which was accepted by one of their *campesino*/artisanal fisher members.

Issues of Equity and Sustainability. An examination of the social processes that have accompanied the expansion of shrimp mariculture raises significant questions regarding the distribution of costs and benefits and the extent to which this latest nontraditional commodity will enhance the income of people in the region. The earlier discussion of land tenure illustrates the unequal distribution of landholdings among shrimp farms. Habitat destruction and over-collection have exacerbated the problem of acquiring naturally occurring post-larval and juvenile shrimp that are used to seed ponds—thereby raising costs, reducing profits, and forcing large companies to import seed stock from

Miami and Panama or to attempt to establish their own nurseries (Castañeda and Matamoros 1990). A continued shortage in post-larval shrimp would most likely not only enhance the internationalization of the industry through increased dependence on imported seed stock but would also exacerbate the already unequal land distribution pattern. As has been the case, it is probable that only the large firms could afford to import stock in order to continue operation. This in turn may lead to increased expansion of these firms and to an even more skewed distribution of landholdings among shrimp farms.

Finally, the shrimp industry is vulnerable to the same international forces that affected other exports from the region. Factors such as reduced demand, falling prices, reduction in export quotas, and import restrictions due to higher quality control regulations which had significant effects on the cattle boom are liable to influence the shrimp industry as well.

The Melon Boom

The most important of the other nontraditionals promoted in the south are melons—identified by USAID as being among the most competitive crops and able to generate the greatest employment and domestic income per hectare (USAID 1990: 16). Eighty percent of Honduran melon production takes place in the southern region, principally in the departments of Choluteca and Valle (LACR 1989).

As part of its efforts to diversify production beyond bananas on the North Coast, the United Fruit Company began a melon export operation in the south in the mid-1970s, through a subsidiary, *Productos Acuáticos y Terrestres, S.A.* (PATSA), which contracted with small and medium growers for rights to their melon crops. By the early 1980s other producers began entering the market. Braced by stable prices and government incentives, growers aimed to expand the area under irrigation (more than 85 percent of melon fields are irrigated), improve the nutritional levels of their products, and expand distribution in US markets during the winter season (LACR 1989c). Between 1985 and 1989 the number of 15-kilo boxes of melons exported from Honduras rose from 600,000 to 2.4 million (valued at US$8.5 million in 1989). By the 1989-90 growing season 4,500 hectares were in melon cultivation approximately half of which was being cultivated by small farmers (USAID 1990). Industry sources project an annual expansion rate of 600 to 700 hectares before reaching 6,000 hectares—the amount of land targeted for eventual conversion (Murray 1991).

The social and environmental consequences of the melon boom are as yet unclear, but several familiar predicaments are emerging. Boyer's budget studies of PATSA members in the late 1970s showed that the small and medium producers were barely breaking even (1982). Murray's (1991) more recent study of the southern Honduran melon industry, suggests that among the effects of expanded melon production are an escalation in the rate of land concentration (part of a shift toward greater control of melon production by large and transnational operations) and the concomitant elimination of independent small and medium producers. These results call into question the claim that this latest wave of nontraditionals will support small farmers and encourage equitable growth.

There is mounting evidence, as well, that melon production is exacerbating or recreating many of the same ecological disasters associated with the earlier production of cotton in the region. To control worsening pest problems, caused especially by aphid-born pest viruses, white flies, and leaf-miners, melon growers employ the same pest control strategy used in cotton—applying toxic chemicals on a calendar schedule. The effect has been the emergence of the familiar "pesticide treadmill" brought about as farmers escalate their use of chemical inputs in response to the increased resistance of pests to chemical strategies. Despite these efforts, melon farmers lost an estimated 10 percent of their melon crop to pests in the 1988–1989 season and more than 50 percent during 1989–1990 (Murray 1991). Although these losses were partially off-set by high international prices, these occurrences call into question the long-term economic and ecological sustainability of current agricultural practices (Meckenstock et al. 1991).

Other vital concerns are related to the social and environmental consequences of expanded irrigation for melon production. Water for drinking, bathing, as well as for irrigation, is becoming more scarce in the region. The degree to which capturing larger amounts of water for irrigation is likely to affect access to water, as well as access to land that is suitable for irrigation, for poor farmers is unknown. There is no institutional regulation of well drilling and anyone with enough money to excavate a well can do so—and is doing so in both rural and urban areas. For example, between 1987 and 1990, the regional office of the Central Bank was compelled to drill three wells, each one to a greater depth, because the previous well had gone dry (Castañeda and Matamoros 1990).

The lack of adequate surface water to meet increasing demands, several years of drought, extensive deforestation, and increased areas under irrigation appear to be depleting the coastal aquifer. Instead of contributing fresh water to the Gulf of Fonseca the aquifer is now receiving Gulf water and farmers with irrigation wells are encountering salty or brackish water for the first time. Reduction in the size of the coastal aquifer coupled with contamination from leached pesticides is already affecting water quality, human health, agriculture, and the shrimp industry (Castañeda and Matamoros 1990).

Conclusions

The escalation of capitalist agriculture in the south after World War II had far reaching effects on the allocation and the distribution of land, and in a context of population growth, decreased the availability of land for most people in the region. The expansion took place on large holdings and resulted in the reallocation of land from forest, fallow, or growing food crops to production of export crops and livestock. Small farmers continued to produce most food crops but could not keep pace with growing populations and per capita production of food crops dropped drastically, along with a general decline in nutritional status.

Enlargement of holdings for export production and population increase, combined to concentrate landholdings, especially in excluding small farmers for whom land became increasingly scarce. Nevertheless, as will be shown in subsequent chapters, southern families sought to increase household income in many ways including augmenting agricultural production in marginal areas at the cost of increased deforestation and erosion. Through the 1980s, southern Honduras remained distinguished by a high degree of dependence on the agricultural sector, concentrated landholdings which excluded most of the regions farmers, an expanding population despite continued out-migration, extremely low per capita income, and high levels of unemployment and malnutrition.

Integration into the world system through the expansion of capitalist agriculture, exacerbated by regional population growth, resulted in land, labor, and food scarcity in the south. The concentration of landholdings and population pressure combined to exclude increasing numbers of small farmers from access to land without giving them adequate wage labor alternatives. Crops grown for export removed the means of production from the hands of a large number of households and reduced the availability of food grains for those people. The failures of export crops due to recurring "bust" cycles decreased employment opportunities. These factors combined to result in extensive malnutrition in the region. Without sufficient resources to grow or to buy food, food scarcity and related undernutrition resulted. However, resources were not scarce for everyone; land was not scarce for large landowners, wage labor was not needed by those with other alternatives.

In response to market forces, regional, national, and international elites and transnational corporations shifted from the production of one export commodity to another—cotton, sugar, cattle—with little regard for the environmental and social costs, and embraced the emerging opportunities presented by the new set of nontraditional exports—opportunities brought about, in part, by previous crises in the export sector. There is evidence, as well, that large scale operations predict and integrate crises, generating environmental and social problems into their investment and production strategies— investing profits and expanding into new regions before anticipated ecological imbalances, economic decline, and social unrest become too severe.[9]

The recent promotion of shrimp mariculture, in particular, has extended the catastrophic social and ecological processes that took place primarily in highland, foothill, and lowland areas, to coastal zones. The privatization of state lands attended by diminished available resources for communal use, escalating land values, enclosure movements supported by force and accompanied by rural displacement and violence, are not new phenomena in the south. Neither are massive deforestation and destruction of habitats emanating from the expansion of capitalist agriculture. A significant difference, however, between previous capitalist growth and the promotion of the new nontraditionals, is that the local people now being displaced embody a semi-proletarianized group that had been partially dispossessed because of earlier capitalist agricultural expansion. At the start of the shrimp and melon booms, the household economic survival strategies of these families consisted of multiple income generating activities including the exploitation of coastal resources, petty commodity production, part-time wage work, and cyclical migration. The pattern in the growth of the new nontraditionals has diminished the economic resource base of these families still further. With record unemployment and underemployment in the country, there is little hope that those displaced as a result of the expansion of shrimp and melon farms can be absorbed into other sectors of the economy.

Notes

1. These funds were released to the new Honduran government that took office in January 1990.

2. In July 1991 the Honduran Central Bank reached an agreement with the IMF which paved the way for an influx of US$1.8 billion worth of external finance over a three year period: US$300 million in 1991, US$70 million in 1992, and 750 million in 1993 (LARR 1991a: 5).

3. Prior to the devaluation of 1990, 1 Lempira = US$.50. The unit of land used most by farmers is the *manzana*: 1 *manzana* = .69 hectares.

4. Begun in the 1960s as a response to peasant demands for land, most reform communities were formed on expropriated or national lands and were administered by the *Instituto Nacional Agrícola* (INA) which works closely with international agencies. Boyer (1982) is especially helpful in understanding the human and social consequences of the link that developed between agrarian reform communities, loan capital, and reform bureaucracies. He concluded that by limiting loan capital to a few favored groups and a few favored export crops, the agrarian reform program met only one of its major goals—expanding capitalist agriculture—while it failed to meet its other major goal of redistribution.

5. By the mid-1970s, the agrarian reform process slowed decisively throughout Honduras due to a marked lack of political will on the part of various Honduran governments, significant resistance by large ranchers and farmers, unwieldy bureaucratic requirements, and corruption at INA (Ruhl 1984, 1987; CEDOH 1988). As an alternative, the government, with the assistance of USAID, promoted a land titling program aimed at providing legal claim to lands which farmers had cultivated for years—presumably as a way to facilitate access to credit. The program failed to address one of the most serious problems in rural areas—the growing number of peasant families with no land at all.

An adjunct to the land-titling program was a mortgage financing project initiated in 1983, in which USAID provided the Central Bank with a pool of local currency that was to be made available to private banks for financing mortgages for agricultural land purchases. The stated purpose of the project was to help landpoor farmers expand their landholdings. Evaluators found, however, that because the donors did not adequately monitor the project, a significant portion of the benefits flowed to the middle class—although some landless and near landless peasants did acquire land (Forster 1992: 575). Only two banks—out of 15 eligible private banks—chose to participate in the program, and of these, one issued 90 of the 93 loans. In general, the bank provided funds to its regular clientele—the urban middle class and medium and large farmers with a credit history at the bank. An additional constraint on small farmers was the tradition of Honduran banks to protect themselves. According to law, they can lend for only 60 percent of the value of the property being purchased, although the full value of the land serves as collateral. After a 10 percent down payment, borrowers must provide additional collateral to cover the full purchase price (e.g., a savings account at the bank, crops, or a guarantor). These harsh requirements eliminated most poor farmers (Forster 1992: 575).

With no acceptable options, peasant organizations responded with direct action in the form of local and nationally coordinated land occupations designed to pressure the government into implementing its own agrarian reform law. These occupations continued into 1992, after the passage of the so-called agricultural modernization law—a law which, in effect, closes the books on the agrarian reform by ending the setting aside of land for redistribution and downgrading INA to the status of a land titling agency. The only remaining incentive to redistribution is penalizing the practice of holding land idle through taxation (LARR 1992b).

6. The recent expansion of shrimp farms is only one of the causes of the destruction of mangrove ecosystems. Approximately 500 salt making operations use mangrove wood as fuel in ovens used in the salt extraction industry: in 1988 these businesses used 50,000 m of mangrove and other wood. In the same year eight tanneries purchased 324 metric tons (approximately 1,500 trees) of mangrove wood for use in that industry. Finally, mangroves also provide an important source of domestic fuelwood, estimated to be 24,000 m per year (SECPLAN/USAID 1989). Destruction of habitats, blocking of estuaries, and rechanneling of rivers associated with the above efforts also encourage ecological imbalances and destruction of other flora and fauna.

7. Other estimates of direct employment range from 1.2 jobs per hectare (including full time farms and packing plants) to 5 jobs per hectare (on small farms) and from 3,000–8,900 full time jobs.

8. In 1991, CODDEFFAGOLF received a grant of $100,000 from the Inter-American Foundation to finance and teach sustainable agricultural techniques to approximately 100 peasant families and to help 331 coastal families establish modern salt-evaporation and aquaculture ponds. CODDEFFAGOLF has also been successful in obtaining funding for conservation projects from the World Wildlife Fund.

9. For example, in Mexico, melon exporters acknowledged integrating the anticipation of escalating pest, economic, and social problems into their investment strategy, planning for a seven year production cycle in a particular community or agricultural zone before moving to new areas (López 1990). A spokesperson for the melon industry in Choluteca who was interviewed by Douglas Murray in 1990 revealed that large growers in Honduras had learned from the experience of their counterparts in Mexico. Lamenting the possibility of a continued crisis in melons, he remarked: "Like the Mexicans, we may have to move if the problem isn't solved. We are currently looking into some disease-free zones in Nicaragua. If we have to, we can move these packing sheds and equipment in two days" (cited in Murray 1991:26).

5

Local Level Responses
to Agrarian Transformation:
The Municipality and the Community

Previous chapters examined the interactions among demography, the physical and natural environment, historical and contemporary attempts at development, and associated human and environmental consequences—at the level of the region. This chapter begins the focus on more micro-levels (the municipality, community, household, and individual) and links important processes and mechanisms occurring at these levels to those of the region. Earlier chapters described the entrances of, and alliances among, powerful international and national actors and their profound influences in the south. This chapter and the next, analyze the interrelations between these agents and local people, especially regarding connections between the political and the household economies, the roles of external actors in affecting the available options in the use of natural resources, and the subsequent decisions of local resource managers. In carrying through the political ecology perspective, described in Chapter 1, analysis does not stop at examining the profound effects of these powerful groups at the local level, but proceeds to examine the heterogeneity in local level responses, and how, in turn, the coping strategies that were created and implemented are affecting more macro levels—the region and the nation. Such strategies include changes in the household economy, modifications in the prevailing agricultural systems, and alterations in the use of natural resources in general. Integrated into the analysis, as well, are important demographic dynamics taking place at the local level. This chapter centers on the municipality and the community, and provides the framework with which to comprehend more fully, the following chapter which emphasizes social differentiation, variations in the household economy, and the relationship between modified agricultural systems and environmental destruction. This chapter introduces two highland communities located in the municipality of Pespire and describes diversity in local level social, economic, and ecological patterns.[1] First, land use, land tenure, and demographic data from the municipal level are related to the regional context examined earlier. This is followed by ethnographic descriptions of the highland communities of San Esteban and

Oroquina that concentrate on commonality and variation between and within the communities and on the effects in these villages of the various crises of the 1980s.[2] Like most other communities in Pespire, neither San Esteban nor Oroquina were in the immediate route of the military actions that took place in southern Honduras during the decade. Although occasional patrols of Honduran soldiers passed through or near the villages, Contra commanders or bands of Contras did not camp close-by. Neither were San Esteban or Oroquina the focus of any large scale economic development efforts by USAID, by other multilateral or bilateral assistance agencies, by the Honduran government, or by any of the many NGOs with projects in the region. Nevertheless, the villages were profoundly affected by the international and national decisions that were made during the period.

The Municipality of Pespire

The municipality of Pespire is located in the Department of Choluteca in the north-central portion of the southern region, and is comprised of a municipal center, also named Pespire, and nine villages, each associated with numerous smaller hamlets. These outlying communities are connected to the town of Pespire by a network of unpaved roads and footpaths. Although limited bus service connecting some of these communities to the municipal center is available, service tends to be erratic, and many communities are difficult to reach, particularly during the rainy season when roads become muddy and fords too deep to cross.

The major paved spur of the Pan American Highway joining the south coast to Tegucigalpa passes through the municipal center, making Pespire accessible to the rest of the country. Either traveling south from Tegucigalpa, or north from Choluteca, San Lorenzo, or Nacaome, the town of Pespire appears very picturesque. The Spanish colonial architecture, the golden domes of the church, the white washed and brightly painted houses, and the children playing in the broad main street, combine to give Pespire a quaint, small town appearance. There, too, many amenities of modern life are available; electricity, a water system, telegraph, medical clinic, national and municipal offices, an elementary school, numerous general stores (*pulperias*), and several modern houses. The town center contrasts sharply with the outlying villages and hamlets where most of these amenities are not available and where people appear more indigent.

The existing town center lies a few hundred meters north of a probable indigenous settlement that was situated at the confluence of the Nacaome and Sacamil Rivers. This settlement became the site of a colonial *reducción*; the census of 1801 mentions a population of 24 ladino families and 37 Indians living there at that time. No *Spanish* are mentioned in the census and it is likely that Pespire was never the site of large landholding estates (Stonich 1986).

Since 1950, the population of Pespire did not grow as quickly as the population of the south, nor of Honduras as a whole, because of high rates of out-migration and infant mortality. Despite these constraints, rural population densities in Pespire continued to be greater than the regional and national averages: increasing from 35 inhabitants per square kilometer in 1950, to 54 inhabitants per square kilometer by 1974 before reaching 74.3 people per square kilometer by 1988 (computed from DGECH 1981; SECPLAN 1989).

Pespire declined in importance as an urban center after 1950, along with the emergence of Choluteca, San Lorenzo, and Nacaome, as the primary regional centers (Stonich 1986). Between 1974 and 1988, the annual urban growth rate of Pespire was 2.4 percent, compared to 5.4 percent for Choluteca, and 5.6 percent for San Lorenzo. By 1988, the total population of Pespire was approximately 25,000 with 2,600 living in the town center. Pespire is more rural than the region as a whole: between 1950 and 1988, the urban population stayed at approximately 10 percent of the total population, with urban dwellers limited to the municipal center. In town, people make their living working in the municipal and national government agencies, various services, and businesses. Because of the relatively easy trip by bus to Tegucigalpa, members of many families live and work in Tegucigalpa during the week, and return to Pespire on weekends or less frequently. Most of the Pespire's population, however, live in the outlying communities and are directly or indirectly dependent, at least to a certain extent, on agricultural production for their livelihoods.

Although 96 percent of Pespire is classified as subtropical humid forest, considerable environmental diversity exists within the municipality (CRIES 1984). Elevations extend from 0 to 1500 meters; average annual precipitation varies from 500 to 2500 millimeters; and average annual temperatures range from 22 degrees Centigrade to 30 degrees Centigrade. In general, temperature and precipitation decline with increasing altitude (CRIES 1984, unpublished data).

The western portion of Pespire is located in the foothills, where the generally flat coastal plain gives rise to the highlands that encompass the eastern portion of the municipality. The foothills themselves are extremely steep and the principal types of agriculture practiced there, as well as in the rocky soils of the highlands, are shifting cultivation (both slash-and-burn and slash-and-mulch systems) and livestock raising. There are a few fertile river bottoms where plow agriculture is possible, but these account for less than 20 percent of agricultural land in the municipality (CRIES 1984, unpublished data).

As in the rest of the south, a distinct rainy season occurs from May to November alternating with a five month dry season during which almost no rain may fall. In the middle of the rainy season there is usually a period that is quite dry (*la canícula*). Although, farmers admit that this dry period does not always happen and that its onset and duration are quite variable, it generally takes place sometime between July 15 and August 15. It is during this time that maturation of the first corn crop takes place and prolongation of this dry period can result in substantial crop loss. Although average yearly precipitation is approximately 1,600 millimeters, there is considerable variation in the amount of rain that falls from year to year, as well as from month to month. This is illustrated, in part, by the pattern of rainfall in 1982; after slightly above average precipitation rates for the beginning of the year, rainfall reached a near record high of 612 millimeters in May, before falling to below average amounts for the rest of the year. The May storms brought widespread flooding, dangerous

landslides, extensive erosion, and considerable human disease and death to
the entire Pacific coast of Honduras, Nicaragua, and El Salvador. The period
of drought that began in 1982, marked the beginning of several consecutive
years of desiccation that lasted through most of the decade. Drought conditions were especially severe in 1986 and led to the creation of a special program of food aid and other assistance for the region. Southern farmers suffered not only the devastating effects brought about because of depressed
yields of corn (a major food crop) but the loss of more drought tolerant sorghum (also a major human food crop) as well. In fact, 1986 was the first year in
the memories of area farmers, and of agricultural extensionists alike, during
which almost the entire sorghum crop was lost because of insufficient precipitation. This prolonged period of drought was broken temporarily during the
winter of 1988 when, once again, extremely heavy rainfall brought widespread floods and landslides.

Pespire in the Regional Context

Many of the changes in the allocation and the distribution of farmland and in
land tenure that were manifest at the regional level also were evident in
Pespire. Changes in land use between 1952 and 1974 in Pespire and in the
southern region are compared in Figure 5.1. In Pespire declines occurred in the
total area planted in annual crops (from 18 percent to 14 percent), permanent
crops (from 2 percent to 1 percent), land in fallow (from 30 percent to 5 percent), and forest land (from 26 percent to 22 percent), while during the same
period pasture land increased significantly (from 22 percent to 57 percent).

The implications of these ecological transformations on the production of
food crops and cattle is suggested by Figure 5.2 which includes changes in the
number of farms, area of farmland, and total production of corn, beans, sorghum, and cattle between 1952 and 1974. The only food grain to show an increase in any category was corn planted in the first planting season (*primera*) (a
small increase of 2.4 percent in the total number of farms planting corn). Even
in this case, the total amount of farmland planted in corn decreased 16 percent
and total corn production decreased 4.4 percent during the period. With the
exception of beans sown in the first planting season, for which production remained constant, the area in cultivation as well as total production decreased
for all other food grains. The production of food grains stands in marked contrast to cattle production; although 11 percent fewer farms raised cattle in 1974
than in 1952 there was a 154 percent increase in the area in pasture and a 68
percent boost in the number of cattle raised. As was true at the regional level,
cattle production was closely associated with the size of farms (Stonich 1989).

Table 5.1 summarizes the percentage of land in cultivation of annual crops
(for the most part food crops and cotton) in each farm size category in Pespire
in 1974. It reveals a negative relationship between the size of farms and the
percentage of farmland under cultivation: 97 percent of farmland on farms of
less than one hectare was in cultivation, 56 percent of land on farms of from
one to five hectares was being cultivated, and so on—as the size of farms increased the percentage of farmland in cultivation decreased. (The augmented

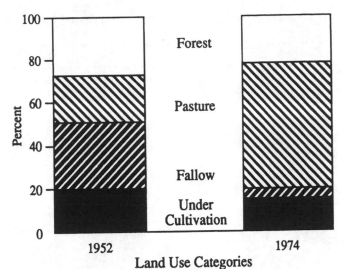

FIGURE 5.1 Changes in Land Use in Pespire: 1952 and 1974.
Source: DGECH 1954, 1976.

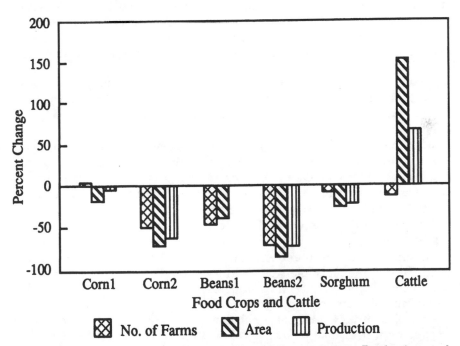

FIGURE 5.2 Percent Changes in the Number of Farms, Area in Production, and
Total Production of Corn (first and second planting), Beans (first and second planting), Sorghum, and Cattle in Pespire: 1952–1974. Source: DGECH 1954, 1976.

TABLE 5.1 Percentage of Land in Cultivation by Farm Size Category in Pespire: 1974

Farm Size Category[a]	Area[a]	% in Cultivation
<1	242	97.0
≥1, <5	1,734	55.8
≥5, <10	1,920	23.6
≥10, <20	2,367	15.1
≥20, <50	3,798	9.1
≥50, <100	2,481	5.6
≥100, <200	2,194	5.7
≥200, <500	1,962	2.6
≥500, <1000	1,162	3.5
≥1000, <2500	1,523	16.1[b]

[a] in hectares.
[b] cotton.

Source: DGECH 1976.

percentage of land being cultivated on farms of more than 1,000 hectares was due to the cultivation of cotton on such farms, most of which were located in the limited lowland areas). Although, as these data show, small farmers used their land most intensively to produce food crops, per capita production of human food grains (corn, beans, and sorghum) declined significantly between 1952 and 1974: corn and sorghum by 46 percent and beans by 96 percent (Stonich 1986:175).

In 1974 as in 1952 the majority of farmers in Pespire remained smallholders. During the 22 year period, however, the distribution of farmland became more concentrated. By 1974, farms of less than ten hectares comprised 80 percent of all farms in Pespire but controlled less than 20 percent of the total farmland, while farms of more than 50 hectares (3.5 percent of all farms, 61 landowners) owned close to one half of all the land in the municipality. Land concentration would appear more skewed if the growing number of landless families were included in these data. Increased scarcity in land is also suggested by the increase in the number of rented farms during the period—from 10.45 percent to 35.6 percent of all farms in the municipality (computed from DGECH 1954, 1976).

In summary, lacking the extensive lowlands suitable for large scale cultivation of cotton and sugar, the capitalization of agriculture in Pespire focused on augmented cattle production. The aggregate data suggest that this process contributed to further land concentration and decreased the production of food grains during a period of population increase which together resulted in decreases in per capita production of food crops. The mechanisms through which this occurred in highland communities and the human and environmental consequences of the processes involved are the focus of much of the remainder of this volume.

Two Highland Communities

The adjacent highland villages of Oroquina and San Esteban lie approximately 25 kilometers southeast of the town of Pespire. Much of the narrow unpaved road leading to these two highland villages follows the course of the old footpath that joins Pespire with the town center of the neighboring municipality of Choluteca. During the dry season the road is hot and dusty and crosses numerous sections of exposed limestone bedrock. During the rainy season, the myriad potholes become hidden traps, and the bedrock is transformed into a slippery slide into swollen muddy streams. There are no bridges over the numerous streams, making vehicular travel during the rainy season difficult, even impossible at times. The present road was built in the early 1970s, as part of national attempts at regional integration and development. The main section of the road by-passed Oroquina and San Esteban, but local leaders in Oroquina organized the construction of a spur to their community that was completed in 1975. Shortly thereafter, three local residents with financial help from family members living outside the community, initiated bus service to Oroquina and on to San Esteban. They bought two mini-buses and organized other men within the community into road crews. The portion of the road to Oroquina remained passable throughout the 1980s, but by the early 1980s the section to San Esteban was in serious disrepair. In 1982, regular bus service to San Esteban was discontinued, although residents could walk to Oroquina to catch one of the four daily buses to and from Pespire. Although buses were available, the bus fare of approximately US$2 was equivalent to more than a full day's pay as an agricultural laborer, and thus prohibitive for most people except on unusual occasions.

Much of the municipality's environmental, agricultural, and socioeconomic diversity is visible as the road from Pespire climbs to Oroquina, San Esteban, and beyond to Nicaragua. The intersection of the road with the paved highway that joins Tegucigalpa and Choluteca lies south of the Pespire town center. The section of the road nearest the municipal center is in best repair and people and houses along that segment of the road look the most prosperous: houses are made of adobe and painted pink or blue (identifying the residents, respectively, as members of the Liberal or National political party); rolling fields are planted with corn and sorghum; children and adolescents walk along the road carrying notebooks to and from school.

The road narrows as it climbs, passing through smaller villages where houses are composed of a greater variety of building materials; wattle and daub and sticks begin to replace the previously dominant adobe structures. About 10 kilometers from Pespire the road divides at an enormous *ceiba* tree, the main road continuing east to Choluteca, the spur following the gully to Oroquina. The site is a meeting place for peasants dressed in their best clothes who sit in the shade of the tree waiting for buses to Pespire or to hinterland villages. Women carry baskets of fruit, vegetables, eggs, and hard pastries to sell or to give as gifts to relatives. Men carry straw cases that they toss to the top of the bus and tie to the crowded mini-buses. For some, the luxury of a trip to

House of a landless peasant family, made of *estacon* with thatch roof, Pespire highlands. The various measures of wealth calculated for this household were very low.

town on the bus is motivated only by the necessity of visiting the doctor at the health center in Pespire.

Introduction to Oroquina

From this intersection, the road parallels a deep gully into the village of Oroquina. The road suddenly curves and becomes very steep before leveling off at the entrance to the village where it is bordered by large mango and *ceiba* trees. Beyond the trees, the road is flanked on one side by a rocky stream where groups of women and girls wash clothes and gather water. On the other side, white, pink, and blue adobe houses border the way. People wave and call out, frequently asking for a ride to the village center a steep kilometer away. Abruptly, the road ascends to the center of town. Built in the Spanish colonial style, a central square is surrounded by a Catholic Church, an elementary school (grades one through three), a health center, and numerous household residences. Fanning out from the central square are foot paths that lead to the majority of houses. These footpaths also lead farmers to their corn and sorghum fields (*milpas*) which are located from .25 kilometers to three kilometers from their residences.

Most houses are located on *ejidal* land managed by the community. The use rights to this land can be inherited and families with such rights pay an annual tax to the municipality. Although houses range in size from one to five rooms, most houses (80 percent) are one to three room structures. A majority of dwellings (56 percent) are constructed of *estacón* (wood stakes set closely together in vertical columns with no caulking for insulation) or adobe (37 percent). Roofs are made of tile and floors are usually dirt, although a few (6 percent) adobe houses have cement floors. Adobe structures predominate near the more affluent village center and along the major road while *estacón* dwellings are more numerous as distance from the town center increases.

There is no electricity in the village although the owner of the largest general store purchased a generator which he uses to power a motorized mill to grind corn and sorghum. A water system financed with international aid funds was installed in 1975. The water project was initiated by local leaders who successfully solicited funds and technical assistance and who subsequently organized the local labor for the building of the system. All those who worked on the project and their families (77 percent of all households) have water piped to their houses. There is also a public faucet which is used by a few families (8 percent); the remaining villagers (12 percent) draw water from the river that flows through the community. Most households (70 percent) do not have latrines, and of those who do, only very few (3 percent) are in working order.

Until 1989, the village health center was maintained by a practical nurse who was paid by the Honduran government. By the early 1980s, however, all medicines were in short supply and the nurse referred most patients to the center in Pespire where drugs and a medical doctor were available. By the late 1980s in the context of Honduras' worsening fiscal crisis and concomitant austerity measures, the center was closed.

Throughout the 1980s, the primary school had one teacher and approximately 100 students. Students who wished to finish elementary school were forced to commute to Pespire provided that their families could afford the costs of transportation, lodging, fees, uniforms, books, and other supplies.

In the early 1980s, there were five general stores in Oroquina that sold a variety of goods such as coffee, sugar, rice, beans, eggs, corn, medicines, and other assorted products. The four smaller store owners purchased their stock from a peddlar who arrived from Tegucigalpa in his truck every three or four months. The owner of the largest store traveled to Tegucigalpa by bus to replenish his stock. By the late 1980s only two stores remained in business (the largest and one of the smallest) and the variety of good carried by both was significantly less than it had been in the early 1980s. By 1987 the peddlar had stopped visiting the village (he had gone out of business) and the smaller shop owners had no other viable way to procure goods for sale in the village. The owner of the largest store purchased a truck and was able to continue business although on a more limited basis. When asked why the array of goods he sold had declined despite the fact that he now had his own truck with which to transport goods to his store, the owner said that it was because villagers could no longer afford to buy the "luxuries" (e.g., canned goods and fresh vegetables) he had stocked earlier. He now limited his stock to "essentials" such as beans, rice, salt, and sugar. The other general store was fairly small and was able to continue selling goods that the owner regularly transported into the community (he drove the daily bus between Pespire and Oroquina) that were augmented monthly by a female relative who brought them from Tegucigalpa. In all, during the 1980s the variety of goods available to community residents declined considerably.

Introduction to San Esteban

The road from Oroquina passes along the edge of the central square before continuing the five kilometers to San Esteban. For approximately two kilometers beyond the central square the road remains level before commencing the steep ascent to San Esteban. Only a handful of motor vehicles attempt the climb. Most are trucks driven by itinerant merchants (who are called *coyotes*) who contract with farmers to buy grain, avocados, lemons, and other fruit for resale in Tegucigalpa, San Pedro Sula, or the north coast. The road winds along the edge of mountain cliffs flanked by small houses built primarily of sticks held together by woven ropes. Unlike Oroquina, San Esteban has no large town center, church, or health center. The primary school and the adobe structure housing the village agricultural and marketing cooperative are main focal points at which people gather. In general, houses are more dispersed in San Esteban than in Oroquina and families live closer to their fields.

As in Oroquina, houses range from one to five rooms in size with most dwellings (82 percent) having one to three rooms. House types, however are more evenly divided between *estacón* (47 percent) and adobe (44 percent). Most roofs are made of tile (96 percent), but a few are made of straw and palm leaves. All households have dirt floors.

Fewer services and other amenities are available in San Esteban. Like Oroquina, there is no electricity. There is a village water system, less extensive than the one in Oroquina, from which 53 percent of the households get their water. The remaining households, for the most part those more remotely located, access water from a variety of sources; mountain springs (24 percent), public wells (17 percent), private wells (1.3 percent), and the river (1.3 percent). Like Oroquina, an insignificant number of households (3 percent) have a usable latrine.

Because there is no health center, people who need medical care must walk five kilometers down the mountain to Oroquina for care (while that center remained in operation) or board a bus for Pespire. The school has one teacher for approximately 100 students and although it includes grades one through three, most students attend for only one or two years. Through the 1980s, only a few students from the community went on to attend school in Pespire. These students lived with relatives and occasionally returned home.

In the early 1980s there was one small general store in San Esteban which stocked approximately 50 percent of the variety of goods available in the smallest store in Oroquina and which sold goods at slightly higher prices. By 1985 this store was no longer in business and from then on there was no general store in the village.

The History of the Two Communities

The initial settlement of San Esteban and Oroquina has not been well documented; therefore, it is difficult to assess how and to what extent current settlement, socioeconomic, and agricultural patterns have their roots in that time period. Interviews with local residents suggest, however, that the period of initial settlement was basic in the formation of present socioeconomic and land use patterns. According to residents, few people lived in the locale of San Esteban prior to 1900 and of the few resident families none had large landholdings. During the late 1920s and 1930s, five families from the neighboring and more densely settled municipality of Choluteca moved into the area, occupied national lands, and established medium size farms. Their descendants maintain that the initial holdings of these families ranged from 5 to 35 hectares. These founding families grew corn, sorghum, and beans, and raised pigs, chickens, and a few cattle. In the 1940s, a group of cattlemen from Choluteca established a large cattle ranch on the steep, now pasture covered mountainside which overlooks the town.

Oroquina was settled at approximately the same time by two families who occupied 35 to 75 hectares of land, much more than the original settlers of San Esteban. While both families initially concentrated on crop and cattle production, one family quickly established a general store and began buying and selling animals and crops grown by local people for sale in Pespire, Choluteca, and Tegucigalpa. These first founding families were joined by other families, and it appears that by the 1930s the patriarchs of most families living in the community were resident. In both San Esteban and Oroquina the descendants of the pioneering families own approximately 80 percent of all land owned by members of each community although the method of acquisition and the patterns of distribution are much different as will be discussed below.

Village Demography

Reliable population data from the two communities are not available prior to 1971. Analysis of unpublished census and survey data compiled since that time, shows that in San Esteban, between 1971 and 1988, the total population grew from 402 to 488 (21 percent) and the number of households increased from approximately 80 to 94 (18 percent). During the same period, the population of Oroquina escalated from 351 to 610 (74 percent) while the number of households increased from 67 to 106 (58 percent). Of considerable importance is the apparent significant rise in population and in the number of households that took place in Oroquina. Interviews conducted in Oroquina in 1989 and 1990 suggests that many of these households are composed of the families of returning migrants who found it increasingly difficult to find work of any kind in other rural or urban areas.

Although household composition varies between and within communities a majority of households are nuclear families (61 percent of all households in San Esteban and 52 percent in Oroquina). Women headed households (14 percent in San Esteban and 19 percent in Oroquina) include those comprised of

grandmothers plus grandchildren, grandmothers plus daughters plus grandchildren, as well as single women and their children. In only one case is there a
single member household consisting of one woman. Households headed by
men comprise 7 percent of the households in San Esteban and 10 percent in
Oroquina. Other forms of households (those consisting of nuclear family plus
grandchildren, joint families, fosterage, and nuclear families plus fosterage)
make up the remaining 19 percent of households in both communities.[3]

Although neolocal residences is preferred, household units do not ordi-
narily define the boundaries of family economic production units. Many nuclear families are joined into extended family units composed of other households both within and outside the community. While more woman headed
units are also major production units, male headed units are invariably tied to
other households *in* the community either to offspring or to siblings.

Traditional peasant social structure in southern Honduras is based on a du-
ality between nuclear residence and extended kin ties (Boyer 1982). Thus,
while there is a desire to establish separate residence there is also a preference
to locate close to one's parents or older siblings. As will become clear in the following chapter, the result of this pattern of residence is that household economies tend to merge not only around productive but reproductive activities as
well; generally around such relations as labor pooling, non-monetized labor
exchange, petty commodity production and peddling, lending money, and the
sharing of food and other expenses. This traditional pattern of social relations
is much more common in San Esteban where the maintenance of strong extended kin ties within the community and with neighboring communities, as a
defense against scarce economic resources is prominent. In contrast, in more
monetized Oroquina there are significantly fewer food and labor sharing networks, and although, previously, non-monetized labor exchanges were the
norm, they generally have been replaced by cash transactions even among relatives and *compadres.*

On the basis of the household census conducted in 1982 and 1983 the mean
ages of female and male household heads did not vary significantly in the two
communities: the mean age of female household heads in San Esteban was
44.14 ± 15.6 years and 42.25 ± 16.5 years in Oroquina. For male household
heads, the mean age in San Esteban was 44.3 ± 13.9 years and in Oroquina
$43.63\pm$ years The more limited census completed in 1990 indicates that by
that time the mean age of female and male householders had risen approximately two years in both communities: in San Esteban to 46.08 ± 14.2 years for
female householders and to 46.02 ± 12.2 years for male householders; in
Oroquina to 44.9 ± 12.6 years for female householders and to 45.79 ± 11.2
years for male householders. One of the most influential factors contributing
to the elevated average age of householders is the accelerated out-migration of
predominately young people to the two major Honduran cities (San Pedro
Sula and Tegucigalpa) and to the Mosquitia.[4] In Oroquina, it appeared that
two things were happening: young people were continuing their substantial
emigration from the community while many older migrants were returning to
their home community.

The mean number of live births to all women householders from both communities was 6.52± (range 1 to 16)—higher than that estimated for the nation at the same time period (6.1) (CELADE/DGECH 1986). For women who were 45 years of age and older (and hence near the end or past their childbearing years), the mean number of live births was 8.32 children—a little below the estimated total fertility rate for the south at that time (Howard-Borjas 1990: 11).

Infant mortality rates (defined as the number of deaths of infants of less than one year per 1,000 live births) were 117 in San Esteban and 101 in Oroquina, higher than the estimated rate for Honduras (79) and the south (89) (Howard-Borjas 1990: 11).

Table 5.2 compares the two communities to each other and to the national average, on a number of characteristics related to the distribution of populations in different age categories.[5] As shown, the dependency ratios in San Esteban (1.21) and in Oroquina (1.37) significantly exceed those of the nation for 1974 (1.03) and 1988 (1.01) indicating that the two rural communities are composed of a higher proportion of economically dependent individuals than is Honduras as a whole. The association (Pearson Correlation Coefficient) between the number of people per household and the dependency ratio is .467 (p<.0001) in San Esteban and 0.492 (p<.0001) in Oroquina suggesting that as households get larger they tend to be composed of a greater proportion of children and the elderly. Twenty-six percent of the households in San Esteban and 28 percent in Oroquina include grandchildren.

In both communities, few of the children of householders who are less than 13 years of age, live away from their parents; however, once children reach 13 years of age, and essentially assume adult economic responsibility, the pattern diverges.[6] In San Esteban, of the 221 children of householders who were greater than 13 years old, 39 percent live in the household of their parents, 30 percent in some other household in the village, and 31 percent outside of the community. In contrast in Oroquina, of the 213 children, 31 percent live with their parents, 21 percent in other households in the community and 48 percent in some other community. The higher percentage of children 13 years of age and older living with their parents in San Esteban is due, in part, to the emerging tendency for married children to live with their parents rather than establish separate households. The dissimilar allocation of children living outside the home in the two communities reflects the enhanced tendency for children from Oroquina to emigrate from the community.

Diet, Nutrition, and Food Security

Despite declines in regional and municipal per capita food grain production, families remained extremely dependent on such grains for both energy and protein: for most households grains constituted the majority of energy and protein intake and consumption of grains was highly correlated with the extent to which household energy needs were satisfied. Table 5.3 compares the national per capita supply of various food groups to diet as determined by FAO food balance sheets (FAO 1984: 103–4); the proportion of total energy in-

TABLE 5.2   Percentage of Population in Separate Age Categories and Associated Dependency Ratios[a]: San Esteban and Oroquina (1982–83) and Honduras (1974 and 1988)

| Age Category | Communities | | | Honduras | |
	San Esteban %	Oroquina %	Both %	1974 %	1988 %
Elderly (≥65 years)	3.6	3.2	3.5	2.8	3.5
Young (<15 years)	51.1	54.6	52.8	48.1	46.8
Working population	45.3	42.2	43.7	49.2	49.7
Total	100	100	100	100	100
Dependency Ratio	1.21	1.37	1.29	1.03	1.01

[a] Dependency Ratio is the ratio of the population defined as dependent (under 15 and over 64 years of age) to the working population (15 through 64 years of age) (UNDP 1990).

Source: For San Esteban and Oroquina, computed from survey data.   For Honduras computed from data published in SECPLAN 1989.

TABLE 5.3 Percentage Contribution of Food Groups to Diet in Honduras

Food Group	National Per Capita Supply[a]	INCAP Rural Area 1969[b]	Dry Season[c]	Rainy Season[c]
Dairy Products	4.5	7.0	1.2	1.3
Eggs	0.5	1.0	0.8	0.6
Meat	2.4	5.0	3.5	2.4
Beef	(1.1)	—	(1.2)	(1.2)
Pork	(0.4)	—	(1.7)	(0.4)
Chicken	(0.5)	—	(0.6)	(0.7)
Beans	4.3	11.0	1.4	1.0
Fruits/Vegetables	8.8	5.0	7.3	6.5
Grains	54.8	50.0	75.3	77.1
Corn	(41.7)	—	(31.8)	(67.2)
Sorghum	(3.1)	—	(37.2)	(1.2)
Rice	(3.6)	—	(3.7)	(4.5)
Wheat	(6.2)	—	(2.6)	(4.2)
Fats	8.5	—	7.2	5.0
Sugar	15.2	8.0	5.5	5.8
Coffee	0.2	—	0.06	0.0

[a] Based on FAO Balance Sheets (1979–81).
[b] INCAP Sample of Rural Areas in Honduras (1969).
[c] Rural Highland Sample (Dry Season 1983 and Rainy Season 1983).

Source: [a] FAO 1984. [b] INCAP 1969. [c] Calculated from household diet study.

take represented by these same food groups as described by INCAP (1969) for *rural* areas of Honduras; and the percentage contribution of these foods as calculated for both communities (once in the dry and again in the rainy season). It is noteworthy that according to the FAO food balance sheets and the INCAP study, grains contributed between 50 percent to 55 percent of energy intake, while in the highland sample a much greater proportion of energy intake was provided by grains: from 75 percent in the dry season to 77 percent in the rainy season. Further, although the proportion of energy intake provided by grains remained similar in both seasons the relative contributions of corn and sorghum differed significantly depending on the season. Consumption of sorghum was greatest in the dry season immediately after the sorghum harvest, although some families reported using sorghum as the primary grain from which they made tortillas for up to nine months of the year. Sorghum use was greater in the highlands among tenant farmers, sharecroppers, and those with less access to land. During drought years when the corn crop was lost, sorghum use for human food increased. In general, sorghum was consumed by the neediest individuals in the population, and by the entire population during times of economic stress (Stonich 1991b).

Foods of animal origin (dairy products, eggs, and meat) provided 13 percent of energy intake (32 percent of protein intake) in the INCAP study and 7.4 percent according to the FAO food balance sheets, but only 5.5 percent of energy intake during the dry season and 4.3 percent during the rainy season according to the highland sample. Although for rural households beef constituted only 1.2 percent of energy intake and 3.4 percent of protein intake, consumption of beef was positively correlated with the amount of land owned, total access to land (both owned and rented), and other measures of wealth including the number of cattle owned by the household (Stonich 1991b). These data provide evidence that the penetration of commercial cattle production into highland areas is having the most direct nutritional effect on wealthier farmers who also tend to be the livestock owners. Analysis of the relationship between the use of particular foods and the degree to which energy and protein needs were met or exceeded by each household shows that the use of grains and the use of animal products are highly correlated with meeting energy needs in both villages studied (Stonich 1991b).

Nutritional Adequacy and Food Security. In the sample of 150 families (75 from San Esteban and 75 from Oroquina) for which dietary data were collected for a one week period during sequential dry and rainy seasons, there were no statistically significant differences, by season, with respect to the mean percentage of household energy and protein needs satisfied nor in the percentage of households that did not meet these requirements. For the dry season the average intake for the entire sample was 114 percent of energy needs and 216 percent of protein needs. These average figures, however, mask the diversity that existed between and within the communities. In San Esteban the mean percentage of energy needs satisfied was 100 percent (range 24 percent to 217 percent) while in Oroquina it was 128 percent (range 24 percent to 343 percent). In San Esteban approximately 57 percent of households did not meet their energy needs in contrast to Oroquina where 37 percent failed to meet their requirements.[7] Analysis of diets showed protein intake tended to surpass protein needs even when household energy intake was unsatisfactory: only 1 percent (1 household) in Oroquina and 4 percent (3 households) in San Esteban did not meet their protein requirements.[8]

Analysis of household food security, measured in terms of the extent to which the *same* household met energy requirements at both points in time, however, demonstrated the considerable nutritional vulnerability of the majority of families. The Pearson correlation coefficient comparing the percentage of energy needs met in the dry season with the percentage of needs met in the rainy season was $r=.421$ ($p<.001$) in San Esteban and $r=.114$ (not statistically significant) in Oroquina. For all households sampled, 27 percent remained unable to satisfy their energy needs, 30 percent sustained their energy requirements at both points of time, and 43 percent fluctuated either from satisfying needs to not satisfying requirements, or the reverse (Stonich 1991b).

Health and Health Care

Until the late 1980s, health care alternatives for villagers included treatment by local injectionists and by the health care practitioner at the Health Center in Oroquina, referral to the municipal Health Center in Pespire which has a physician in residence, private physicians in Nacaome, San Lorenzo, Choluteca, and Tegucigalpa, as well as hospitals in Choluteca and Tegucigalpa. By the early 1980s, the Health Center in Oroquina was short of all kinds of medicines, rarely had any antibiotics to treat the most common illnesses, and functioned primarily as a referral service for seriously ill patients to the clinic in Pespire.

The Health Center in Pespire offers treatment by a physician for infectious diseases, a well baby clinic, prenatal care for mothers, and vaccinations for children. Most of the cost is subsidized by the government; however, by the late 1980s pharmaceuticals were so scarce throughout Honduras that the Health Center was operating basically with no medicine to dispense.[9]

Health care outside the community was inconceivable, or extremely unlikely, for most residents in both communities. For residents in San Esteban, the steep, hot, and dusty (or muddy) walk to Oroquina was extremely difficult for the ill.[10] The cost of the bus trip to and from Pespire, plus the cost of medication (if available) and possibly of a nights stay with relatives or friends, was prohibitive. For these reasons most illnesses are treated at home. Flu, diarrhea, fever, headaches, and malaria are common illnesses throughout the year, with the number of flu cases increasing during the rainy season; while eye infections, bronchitis, and skin rashes are a frequent problem particularly during the dry season.

Extra-Household Social and
Economic Linkages Within the Community

Clubs and Associations are more common in San Esteban than in Oroquina with the exception of *patronatos* (organizations of male householders formed to undertake community projects such as latrine building, road building and maintenance, school construction, and water system upkeep) which exist in both villages. Of greatest importance in San Esteban is the agricultural cooperative to which 57 percent of all households belong. The cooperative in San Esteban is a remnant of a more extensive network of cooperatives begun in the late 1970s. There was an operating cooperative in Oroquina until 1980 when, according to residents, mismanagement resulted in its disbanding. In San Esteban the cooperative operated throughout the 1980s although in an attenuated form. Major functions are to finance sorghum production, buy and sell grain to members, and help in the marketing of sweets made from community grown and processed sugar cane and bread. It also manages a small amount of land which is made available tao members in return for a share of the harvest. Other major voluntary associations in San Esteban included a *Club de Amas de Casa*, an extension service of the Honduran Department of Social Welfare, designed to provide instruction and assistance in animal husbandry, agriculture, health, and nutrition, and a chapter of the Catholic Church sponsored *Club de Amas de Hogar*. Members of both these homemaker clubs consist of women householders.

Although during the 1982 to 1984 study periods several Protestant churches had congregations in foothill communities of Pespire, all families in Oroquina and San Esteban reported that they were Roman Catholic. By 1989, 25 percent of the households surveyed in Oroquina and 15 percent in San Esteban said that they were members of one of several Pentecostal sects.

Agricultural Systems and Use of the Environment

Socioeconomic Constraints:
Land Tenure and Access to Land

Given the regional context, understanding local use of the environment must include a discussion of land tenure as a major constraint. Figure 5.3 summarizes the land tenure arrangements in San Esteban and Oroquina in the period 1982 to 1983.[11] The category of "renters" includes all households who rented land for cash payment, who sharecropped,[12] or who were loaned land. In San Esteban, 5.7 percent of households neither owned nor rented land, 31.4 percent rented land and did not own land, 30 percent both owned and rented land, and 33 percent owned and did not rent land. The distribution of land tenure was quite different in Oroquina where (as in San Esteban) 5.7 percent of households were neither owners nor renters, but 67.2 percent rented land, 13.4 percent both owned and rented land, and 13.4 percent owned land. Thus, in Oroquina land ownership was concentrated in the hands of 27 percent of all households while in San Esteban land was more equitably divided among 63 percent of households.

"Renting" arrangements also differed in the two communities: in San Esteban, 36 percent of households were loaned land by family members, 23 percent sharecropped, and no households rented land for a cash payment. In Oroquina, however, only 19 percent of households were loaned land while 57 percent were sharecroppers, and 4 percent rented land in exchange for a cash payment. Although four households in San Esteban neither owned nor rented

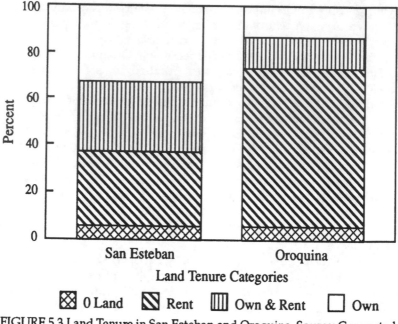

Land Tenure Categories

⊠ 0 Land   ◩ Rent   ▥ Own & Rent   ☐ Own

FIGURE 5.3 Land Tenure in San Esteban and Oroquina. Source: Computed from survey data.

land, all aided a family member in the cultivation of that member's land, for which they were given a share of the crop. In contrast, three of the four non-owners/non-renters in Oroquina relied on wage work as their primary household income generating activity.

The degree of concentration in the size of landholdings, both owned and rented, as well as in the number and percentage of landowning households also varied considerably between the two communities (see Table 5.4). Values for all of the following variables were lower in San Esteban than in Oroquina; the mean amount of land owned per household, the range in the total amount of land owned by households, the mean amount of land rented per household, the mean amount of land to which households had access (the amount owned plus the amount rented), and the range in the total amount of land to which households had access. Thus, even though land was more equitably distributed among households in San Esteban, in total, households there controlled less land.

In both communities the descendants of the initial founding families owned approximately 80 percent of the total land owned by members of the community. However, in San Esteban, the land was distributed among 26 (37 percent) households while in Oroquina, 6 (9 percent) households owned 87 percent of all land owned. The largest landowners, however, did not acquire the majority of their land through inheritance. The 18 households that owned land in Oroquina owned 36 percent more total land than did the 44 landowners in San Esteban. Moreover, more than 70 percent of the land owned by households in Oroquina was purchased rather than inherited. In contrast, in

TABLE 5.4 Household Access to Land: 1982 (Land area in *manzanas*)

	San Esteban	Oroquina
Land Owned:		
Number (%)	44 (63%)	18 (27%)
Mean Owned	4.14 mzs.	16.28 mzs.
Range	.5 to 20 mzs.	.25 to 100 mzs.
Land Rented:		
Number (%)	26 (37%)	49 (73%)
Mean Rented 82	1.26 mzs.	1.48 mzs.
Access to Land:		
(Owned + Rented)		
Mean Access	3.33 mzs.	5.30 mzs.
Range	0 to 20 mzs.	0 to 100 mzs.

Source: Computed from survey data.

San Esteban land was more equally divided between having been purchased (54 percent) and inherited (46 percent).

Patterns in access to land and in land tenure differed between the communities, but neither corresponded closely to regional patterns discerned from the aggregate data sources. There were no large landholders who had access to more than 20 hectares of land in San Esteban or more than 100 hectares in Oroquina. The commonality that crosscut differences between the communities was that most landholdings, either owned or rented, were small, seldom exceeding one *manzana* (.69 hectares) in size. Especially in Oroquina most of these smallholders were not owners but tenants who had little control over long term land management practices and were subject to the decisions of larger holders for access to land. In sum, the majority of cultivation on the steeply sloping hillsides was done by small scale farmers.

Ecological Constraints:
Topography, Rainfall, and Soils

The environmental parameters most important in affecting agricultural production at the regional and municipal levels (topography and climate) are also most relevant at the local level.[13] The two communities can be segregated on the basis of altitude: elevations in Oroquina range from approximately 200 to 450 meters, while altitudes in San Esteban range from 450 to more than 900 meters. Corresponding to these differences in altitude, are differences in mean annual temperature; 24 to 27 degrees Centigrade in Oroquina and 20 to 25 degrees Centigrade in San Esteban. Average annual precipitation also varies between the two communities; 1,500 to 2,000 millimeters in Oroquina and 1,000 to 1,700 millimeters in San Esteban. The patterning of rainfall is a significant constraint with very marked rainy and dry seasons, considerable variability in the amount of rain that falls during the rainy season from year to year and from place to place, and sometimes prolonged dry periods in the middle of the rainy season (*la canícula*). Analyses of soil samples from both communities were judged "remarkably alike" exhibiting a degree of variation no greater than would be expected from soils sampled from the same plot.[14]

Choice of Agricultural Parcels

The rugged, broken topography and erratic rainfall pattern result in considerable micro-ecological and micro-climatic variation within the research areas. Significant variation in the total amount and in the distribution of rainfall often occurs from year to year on the same parcel of land and in the same year on nearby plots. The overall outcome is that the choice of almost any plot for cultivation is a risk in any year. In the past, when land was more available, farmers attempted to minimize this inherent risk by cultivating at least two plots of land in different micro- ecozones—any one of which was perceived to be sufficient to meet the subsistence needs of the household. The recent concentration of landholdings precludes the planting of more than one field plot per household for most farmers, thus leaving families vulnerable to unpredictable yearly fluctuations in rainfall, yields, and total harvested. In 1982/1983, larger landowners in both communities continued the practice of cultivating multiple plots in different micro-ecozones. The largest landowners who allocated portions of their land to sharecroppers, further minimized their risk by contracting to receive a share of the harvest from many plots in various areas. Smallholders and renters had little choice in plots, and most often large holders kept the best land (*tierra buena*) for their own use and rented average land (*tierra regular*) to their landless or landpoor neighbors. Thus, landless and landpoor farmers were limited not only in the quantity but also in the quality of land to which they had access.

Shifting Cultivation Systems

The hillside farmers in San Esteban and Oroquina, like most of the highland cultivators in Southern Honduras, use systems of shifting cultivation that combine both slash-and-burn and slash-and-mulch methods. In the slash-and-burn method, the vegetation on fallow land is cut down and burned at the end of the dry season. The crops that will grow during the rainy season are then planted. While this system is an important component in the overall shifting cultivation system, most often fallow land enters the cropping cycle through a slash-and-mulch method. In this system brush and small trees are cut down during the middle of the rainy season and then allowed to remain in the field to serve as mulch for the crops that must grow up through them. In both San Esteban and Oroquina, land was allowed to lie fallow for from two to five years, after which it could be put back into production, usually using the slash-and-mulch system. Intercropped corn and sorghum are the most important field crops in Oroquina; in San Esteban beans are added to the cropping system. Farmers expect to lose much of their corn crop at least every second or third year and sorghum is intercropped with corn in order to reduce risk. During the years in which drought reduces yields of corn, the much more drought tolerant sorghum can be counted on to produce a harvest.

Cropping Systems in Oroquina. Figure 5.4 shows the most common cropping cycle practiced in Oroquina.[15] A field in secondary growth enters the cropping cycle through one of three slash-and-mulch systems used after the *canícula*. In all three systems, the basic technique is to sow either corn or sorghum then chop down the forest cover leaving the vegetation on the field as mulch. The corn or sorghum then grows in the interstices of the dead and decaying vegetation. The first of these methods, called *socolar*, is used for planting corn in the second growing season (*postrera*). With this method the understory of fallow land is cut and removed, after which the farmers plant corn using a digging stick. After the corn has germinated and grown to a few inches in height the larger trees are felled and left in the field. The corn must then grow through the cut vegetation. The second slash-and-mulch system is used to plant sorghum, and is called the *maicillera*. This system only can be used when adequate precipitation is falling because the sorghum seed is broadcast on the fallow land. The vegetation is then cut and left to serve as a mulch. The third system, *guatera*, is a variant of the second. Sorghum is broadcast later in the season and is sown more densely. Farmers using this last method are interested in using the sorghum for the fodder value of the stalk and leaves rather than for grain.

Farmers report that the slash-and-mulch system allows them to obtain an extra harvest from a piece of land. If they cut the forest and burn the land for sowing in the first season, the *primera*, the land can only be used for two years. The second and third year that a field is in cultivation, preparation is with the more common slash-and-burn method. The field is burned in late March or April, then corn and sorghum are planted. The two grains may be planted in the same hole or in separate holes. Farmers with more available labor may wait a week or two until the corn has germinated and then plant sorghum between the hills of corn.

From an agronomic point of view, this system of intercropping corn and sorghum may seem inefficient because corn and sorghum compete for the same soil nutrients. However, from the farmer's point of view, the system makes a great deal of sense. The corn that is planted must be a rapidly maturing variety that finishes its growth before the onset of the dry season. Corn yields are low largely because the varieties mature in such a short time, but they can sustain the farm household for a few months until the sorghum harvest in December.

This farming system is a compromise between the clear cultural preference for corn, the staple of the peasant diet, and the need for the less desirable but climatically better adapted and more reliably yielding sorghum. When corn is not available, sorghum is substituted in the making of *tortillas* and a number of other foods. The function of the more drought tolerant sorghum as a risk reduction crop is illustrated by its performance in 1982. Corn yields from the *primera* were low because the *canícula* began early. Most of the corn was just maturing when the rains stopped. The prolonged *canícula* resulted in farmers being unable to sow their second crop of corn until quite late—about mid-September. Because the rains lasted only a month, the corn from the second planting did not mature and was almost completely lost. Meanwhile, the sor-

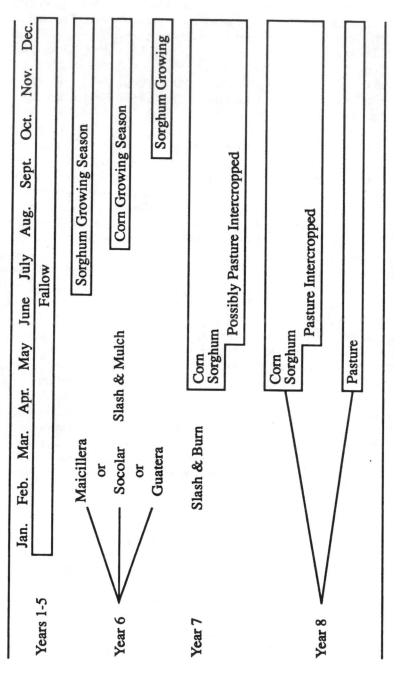

FIGURE 5.4 Cropping Systems, Oroquina. Source: Compiled by author.

ghum tolerated the heavy rains in May and September, the drought in July and August, and the early end to the rainy season in October, and the harvest was near normal.

On the parcels that belong to the larger farmers pasture grasses are often sown along with the corn and sorghum during the third year a field is in cultivation. At the end of the growing season, when the corn and sorghum have been harvested the farmer is left with pasture. In the past the typical pattern was for farmers to graze animals on the pastures for a year or two and then allow natural succession processes to occur. The fields progressed from two or three years in cultivation, to two years in pasture grass, then another three or four years in fallow during which time secondary forest would begin to grow. In the foothill communities of Pespire many of the wealthier farmers, more interested in raising cattle than in growing crops created permanent pastures by weeding and burning fields once a year. These large landowners used landless and landpoor individuals to help them in the conversion from forest and cropland to pasture: land was rented to tenants for a year or two, who then cleared the land, fixed the fences, and in the last year of cultivation sowed pasture grass in the field.[16]

In the early 1980s, this process of permanent conversion was not common in Oroquina and San Esteban but by 1989 was employed regularly in both communities. At the beginning of the decade, the largest cattle owners in Oroquina had a minimum amount of their land (generally their worst land) in permanent pasture. During the dry season when pasture was especially scarce they drove their herds of cattle to the Choluteca lowlands where they were temporarily pastured on large cattle ranches. The large cattle owners from Oroquina did not rent land there but rather paid a set fee per head of cattle per month. The owners, however, increasingly became frustrated with escalating costs, the difficulty of driving cattle through the mountain passes to the lowlands (in part because of augmented U.S. and Honduran military maneuvers and actions in the region), and depressed cattle and beef prices. By 1985 they discontinued the practice of driving their herds to pasture in lowland areas during the dry season. Instead they embraced the practice, formerly found in foothill areas, of transforming additional forest and cropland into permanent pasture. They did not, however, expand the size of their herds, and in fact, herd sizes were virtually the same in 1989/1990 as in 1982/1983. These "rich" peasant farmers with relatively large landholdings believed that during a decade of declining beef prices and severe national economic crisis their most rational economic strategy was to invest in land rather than in cattle.

Cropping Systems in San Esteban. In communities such as San Esteban which are situated at elevations above 400 meters, beans enter the cropping systems. In San Esteban the cultivation cycle begins either with the slash-and-burn system at the start of the rainy season or a slash-and-mulch system after the *canícula*. Corn and beans may be intercropped, beans may be planted alone, or less commonly, sorghum may be planted alone. In the second and third years of cultivation, corn and sorghum are either intercropped (as in Oroquina) or more typically beans, corn, and sorghum are intercropped.

When intercropped, beans are sown up to one week after the corn and sorghum are planted.

Figure 5.5 illustrates the most common 5-year fallow/cropping system practiced in San Esteban. Land enters the cropping cycle in the *postrera* when corn alone, beans alone, or intercropped corn and beans are sown using the slash and mulch system. That is followed by two years of cropping during which time, corn, sorghum, and beans are intercropped in the *primera*. In contrast to Oroquina both slash-and-burn and slash-and-mulch systems are used during this two year period. More than 50 percent of the farmers in San Esteban reported that they do not burn their land—a practice they discontinued in the late 1970s in response to government promotional efforts to encourage farmers to stop burning.

The Technology of Shifting Cultivation Systems. The technology involved in the shifting cultivation systems practiced in San Esteban and Oroquina is fairly simple. The use of fertilizers, (except that provided by burned vegetation and animal manure), plows, and irrigation are virtually absent. Capital investments are minimal. The only tools are *machetes* used for clearing, digging sticks with metal points (*barretas*), gourds used as seed containers, and plastic and string sacks used to transport crops. Seeds are saved from the previous harvest whenever possible. A few farmers own horses or mules which are used to transport grain, but most farmers carry sacks of harvested crops on their back.

The division of labor involved in field crop production is based on sex and age. Women rarely are involved in any phase of the shifting cultivation cycle. By the age 12, adolescent males are given adult responsibility in the fields. For the most part, smallholders maintain their own plots with the help of adolescent sons and male relatives, except when fields are burnt, at which time six to 12 neighbors cooperate in the regulation and containment of the burn. Only a small minority of landowners hire agricultural labor on a regular basis. In San Esteban, none hire labor regularly and in Oroquina such hiring is confined to the two largest landowners. Such holders who generally cultivate bigger parcels of land (usually from three to five hectares as contrasted with the less than one hectare plots of the smallholder producers) hire 10 to 14 men to clear, burn, plant, cultivate, harvest, and process their crops.

Herbicide use was more common in Oroquina where 31 (46 percent) farmers use herbicides for the first weeding. Herbicides were not used as frequently in San Esteban because farmers said that they kill the beans that are part of the cropping system nor were they used in Oroquina by the small group of farmers who continue to intercrop cultigens in addition to corn and sorghum. Herbicides were introduced in Pespire in the early 1970s, and the considerable labor costs necessary to clear weeds by hand have made them an attractive alternative. Few farmers are aware of the name of the herbicide they use although the two most common varieties seem to be 2-4-D and *Herbisol*.

Insecticides are used by 15 (21 percent) households in San Esteban and 25 (37 percent) households in Oroquina. Of these, approximately 70 percent treated seeds before planting—a process that involves nothing more than putting the seeds into the insecticide mixed with a little water and stirring

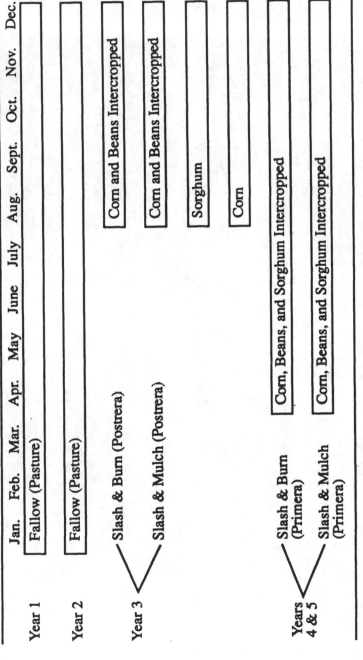

FIGURE 5.5 Cropping Systems, San Esteban. Source: Compiled by author.

Smallholder farmer using a digging stick to plant beans on a steep hillside in Pespire. The plot already has been planted with corn and sorghum.

them around with bare hands. Five percent to 7 percent of farmers in both communities used insecticides on crops growing in the field. *Malathion* and *Dipterex* applied with backpack applicators are the two most commonly used. Few precautions are taken when using insecticides although most farmers seem cognizant in a general way of the their toxic effects.

Cultivation of Other Crops. Although corn, sorghum, and beans (in San Esteban) are the basic subsistence crops, a number of other cultigens are also grown in fields and in house gardens. These provide dietary diversity as well as nutrients not available from basic grains. Squash (*ayote*) is frequently grown as a minor crop. It is planted in May after the corn and sorghum are sown and is ready for harvest in early July, although it can remain on the vine to be harvested as winter squash. Sweet manioc (*yuca*) is also cultivated. It is planted in early May but is not ready for harvest until the end of the year and is usually left in the ground and dug in small quantities when needed. Sugar cane is cultivated by 17 percent of the farmers in San Esteban and 8 percent of the farmers in Oroquina. Plots tend to be small, although one grower in San Esteban has one hectare of irrigated sugar cane planted in a flat area next to a stream. It is planted in June and because it is a perennial is harvested at any time during the year. All the cane is milled locally (San Esteban and Oroquina each have their own mill) and converted into blocks of unrefined brown sugar (*dulce*) which is either consumed at home or sold to stores.

House Gardens and Fruit Trees. House gardens, most often the responsibility of women, provide another source of variety to local diets as well as a source of cash income. Thirty (41 percent) households in San Esteban and 42 (56 percent) households in Oroquina have house gardens. Sweet and hot peppers, tomatoes, *achiote*, squash, beans, *yuca*, cucumbers, onions, and coffee are among the most common crops grown. House gardens have enhanced cash income for a group of women in San Esteban for more than a decade: they began selling the produce that they grew to others in their village in 1979, expanded marketing to include Oroquina in 1980, and enlarged the range of their operations to encompass Pespire and Choluteca in 1982.[17]

Fruit trees, are another important source of food and cash. Sixty percent of all households in both communities have at least some fruit trees which generally are planted close to houses (often as part of house gardens). The most common fruit trees in San Esteban are avocados (57 percent of households), mangos (53 percent), and lemons (33 percent), while mangos (64 percent), papayas (51 percent), bananas (45 percent), lemons (41 percent), and cashews (40 percent) are most plentiful in Oroquina.

Subsistence Crops Are Cash Crops. Despite low average yields and levels of total household production that are insufficient to meet most household nutritional needs, a significant share of basic grains grown by small producers is sold. Although the majority of corn grown in both communities is consumed at home by the producers (e.g., in 1983, 89 percent of corn in San Esteban and 80 percent of corn in Oroquina) the pattern is different for sorghum and beans. In 1983, twenty-five percent of the sorghum and 50 percent of the beans that were grown in San Esteban were marketed, while in Oroquina 63 percent of the sorghum harvest was sold. Most of the sorghum marketed in San Esteban

Woman selling vegetables in the Choluteca market to augment family income.

was sold to the agricultural cooperative while most beans and all the corn were sold to *coyotes*. This pattern is easy to understand since the selling price of beans at the cooperative usually was lower than the prevailing market price, while the selling price of sorghum tended to be higher. About half the farmers selling sorghum to the cooperative reported that they regularly sold their sorghum prior to harvest because they needed cash. With the exception of the largest farmers, most farmers in Oroquina sold their crops to *coyotes* who arrive prior to harvest in order to negotiate selling prices and quantities with farmers (whose home grain supplies tend to be at their lowest) and then return when the harvest is complete. A few small producers sold to local stores and to neighbors as in San Esteban. The largest producers continue to rent trucks with which to haul their harvest to Choluteca or Tegucigalpa in order to obtain the better prices.

Animal Husbandry

Animals provide an important source of income as well as a source of animal protein in the diets of local residents. In 1982/83, the majority of households (63 percent in San Esteban and 58 percent in Oroquina) owned hogs, (including pigs and piglets) and poultry (chickens, hens, and roosters) (54 percent in Esteban and 66 percent in Oroquina). Neither are usually eaten; hogs are normally raised and sold for cash while hens are kept for eggs. Although the same percentage of households (19 percent) owned cattle in both communities, there was a significant difference in the total number of beef cattle owned in each community—102 in Oroquina and 29 in San Esteban. Almost half the beef cattle in Oroquina were owned by one household—that of the largest land-owner while the largest landowner in Oroquina owned the greatest number of hogs and chickens. By 1989/1990, the percentage of households that owned animals of any kind declined in both communities while the range and the total number of animals owned remained approximately the same, suggesting concentration in the ownership of animals.

Animals serve as an important source of cash especially during periods of drought when families are forced to enhance household income because of poor harvests. As drought conditions worsened in 1982 and 1983, the percent-age of households that sold animals in order to obtain cash rose from 20 per-cent to 47 percent in San Esteban and from 48 percent to 72 percent in Oroquina. Except for the few individuals in both communities who regularly bought and sold animals for profit, virtually every seller reported selling ani-mals in order to buy grain—corn or sorghum—for tortillas.

Hunting, Fishing, and Gathering

During the 1982 and 1983 research period, 10 percent of the households in San Esteban and 16 percent in Oroquina reported hunting for food at least one time during the previous year. The most frequently reported game was iguana, but armadillo, rabbits, doves, and opossums were hunted as well. While hunting tended to be an infrequent and primarily male activity, fishing was a more fre-quent and family activity (42 percent of female household heads in San Esteban and 57 percent in Oroquina reported that someone in the family fished on a regular basis). River fish and shellfish caught included sardines (available all year), crayfish (available in July and August) and shrimp (caught in February and March). Approximately 20 percent of the female household heads reported that they gathered plants for food or medicine: nances, guavas, lemons, wild green beans, custard apples, mangos, oranges, and plums.

Fuelwood

Virtually every household used fuelwood for cooking. In the early 1980s, the majority of households (from 70 percent to 80 percent) collected their own fire-wood, while the rest bought fuelwood from someone else in the community who gathered it locally. These figures are consistent with those of Jones (1982) who estimated that 94 percent of all households in the south—compared to 82 percent nationally—used fuelwood to cook and that fuelwood use was ap-proximately 5.83 pounds per person per day. By 1989, the percentage of fami-lies that reported cutting their own firewood locally dropped to 43 percent in San Esteban and 40 percent in Oroquina. Of these 53 percent said that they had cut the firewood from their own property and 47 percent from outside their property. The mean distance traveled to collect firewood was reported as 1.6 kilometers and 85 percent of those who gathered their own firewood reported that they walked more than 2 kilometers in order to do so.

Summary

Although essentially descriptive, the portraits of San Esteban and Oroquina provide evidence of deteriorating economic and human conditions during the 1980s. The increase in the range of the size of landholdings and in the percentage of landless households suggest further concentration of landholdings and exacerbated socioeconomic differentiation. In addition, a decline in the percentage of households that own animals, while the total number of animals remained the same, implies growing inequalities in ownership of animals. The disappearance of the majority of general stores as well as their diminished stock points to the declining purchasing power of most rural residents. A decade characterized by devastating regional drought, deteriorating national and regional economies, and reduced health and social services add to this bleak picture. Moreover, population growth continues to occur in these communities despite significant out-migration. In sum, although considerable diversity exists in the two communities, the 1980s brought about escalated impoverishment and diminished well being.

Knowledge of San Esteban and Oroquina also show that patterns in local level responses to the regional development strategy based on export growth through agricultural diversification varied among communities. In both communities actions did not include direct resistance to incorporation into the capitalist system but rather attempts to shape the way in which each community was articulated into the larger capitalist system. In San Esteban, actions emphasized the formation and expansion of institutions through which the community as a group contends with extra-local forces. To a great extent, the production process could not be tied to single households but rather to relations of nuclear and extended families joined in their capacities to appropriate and allocate land and labor. The proliferation of both male and female dominated voluntary associations, the number and complexity of super-household networks, the increased size of households, and the relatively high percentage of young adults who remained in the community are all evidence of attempts to come to grips with chronic resource scarcity in such a way as to maintain traditional social relations of production and reproduction as much as possible. The transformation of traditional social relations was much more advanced in Oroquina, where community responses involved independent family action to a much greater extent, where people were much more dependent on monetized incomes and on cash transactions, and where merchants and large landowners directly expanded capitalist agriculture into their locale.

Notes

1. This chapter is based on ethnographic and survey research conducted between 1981 and 1991. The chapter does not attempt to capture the range of variation present in municipalities and communities throughout the region during that period. See the following ethnographic descriptions in order to enhance understanding of intra-regional diversity: 1) for the eastern highlands near Nicaragua see Boyer 1982; 2) for the western highlands adjacent to El Salvador see Durham 1979; 3) for lowland and foothill commu-

nities in Pespire see B. DeWalt and K. DeWalt 1982; Fordham, B. DeWalt, K. DeWalt 1985; Thompson, K. DeWalt, B. DeWalt 1985; DeWalt and Stonich 1985.

2. San Esteban and Oroquina are pseudonyms for two communities located in the southern highlands.

3. Until recently the extended family system in which several generations and/or the families of siblings reside together predominated. Data from recent censuses, however, indicate that the nuclear family composed of mother, father, and offsprings is now the norm. It may be deceptive to characterize families as being nuclear or extended because elements of both may be present as is discussed below.

Honduras is no exception to the recent increase in woman-headed households. According to the 1974 census 13% of rural households that included children were headed by women. Migration and abandonment are often the cause of this type of household which are usually organized into some type of extended unit. One common form is to incorporate adult children into the household. Another type is made up of elderly widows. Still another is formed when the eldest daughter in the absence of a male head must provide for an elderly mother or for young brothers not yet old enough to work. In contrast to the large number of female-headed households there is an almost total absence of solitary male households (Zuniga 1978).

Many rural Honduran children experience fosterage at some time. Fosterage is a solution to problems in childrearing that may come about because of general economic conditions, migration, and marital or personal problems. When such problems occur children are sent to live with other family members. Very often small children are sent to live with family members living in rural areas where the costs of childrearing are relatively lower and where children can contribute to the household by performing tasks such as collecting firewood, hauling water, and performing errands. The demand for fosterage, however, seems most often to arise within the family of origin. Elderly family members are a likely choice because of a felt responsibility toward grandchildren (Zuniga 1978). Economic and emotional ties generally are maintained between nonresident parents and their children.

4. Although extensive data on migration behavior were not collected during the 1989/90 research period, local residents concurred that because of the increasingly desperate situation in the highlands younger people and families had intensified their migration out of the area. One item included in the 1989/90 survey administered to male and female householders was to list the three most frequent places to which family members migrated during the previous five years. In order of frequency of responses these were: Tegucigalpa; the Mosquitia (the location of Honduras' last remaining tropical humid forests); and the North Coast (including San Pedro Sula). The significance of this pattern of out-migration will be discussed in subsequent chapters.

5. According to the United Nations, the dependency ratio is the ratio of the population defined to be economically dependent (i.e., under 15 years and over 64 years of age) to the working age population (i.e., 15 through 64 years of age) (UNDP 1990). This definition is problematic when applied to rural populations in Honduras where boys and girls assume adult economic responsibilities and are paid a full wage by the time they are 13 and where the elderly (especially women) perform vital household tasks such as food preparation, child care, and house cleaning. The UN definition is used here to compute dependency ratios because of its comparative value.

6. By the age of 12 or 13, children are well socialized in the roles they are expected to fulfill as adults. Four and five year old children run errands and carry messages. By that age boys begin to learn the mechanics of cultivating the land by following their fathers or older brothers to the fields and learning to swing a machete or to plant with a digging stick. An eight or nine year old boy can perform these tasks efficiently and a ten or 12 year old usually can perform as well as an adult and draws a full wage.

Under their mother's supervision, girls learn the responsibilities associated with domestic work. By seven or eight young girls care for younger siblings, help keep house, feed animals, and develop skill at grinding corn and sorghum for tortillas and preparing meals. A ten or 12 year old girl can manage a household and like her brother readily can fulfill the adult work roles.

7. The results of a nutritional survey conducted by INCAP in 1986 of 1,049 rural Honduran families indicated that 63 percent of the families surveyed consumed less than the recommended amount of calories and of these 50 percent consumed less than 50 percent of the required amount (PAHO 1990). According to the results of the national nutritional survey done in 1987, undernutrition is found in 45 percent of all Honduran children under the age of five. Of these 28 percent of cases (approximately 97,000 children) are severe enough to require hospital treatment even though there are only about 200 hospital beds available for this purpose in the entire country (MSP 1989).

8. This finding is consistent with other studies of the highland south (e.g., Thompson et al. 1985) and with similar studies of other rural Central American populations (e.g., Valverde et al. 1975; Sellers 1984). Whereas earlier research discerned a critical lack of protein in the diets of poor rural populations, actual dietary surveys have tended to find calorie intake more limited than protein.

9. Its aggravated foreign exchange crisis and the imposition of new austerity measures compounded by the demands made by the influx of displaced persons and refugees, the Honduran government found it increasingly difficult to import needed pharmaceuticals and medical supplies. By the late 1980s Honduras experienced chronic and acute shortages of vital pharmaceuticals. Even in the capital city of Tegucigalpa and the industrial city of San Pedro Sula, government health care facilities were critically short of supplies (many of them had been stolen and sold to private dealers) and city pharmacies virtually had empty shelves. Strikes of medical practioners protesting low and unpaid salaries and the lack of supplies ensued despite the efforts of the government to revise the National Health Plan to gear services toward the groups at highest health risk: the rural and urban poor who had the most limited access to health care.

10. According to the Honduran Ministry of Health it takes from one to three hours for 45% of rural Hondurans to reach any health care service (MSP 1989).

11. The figures and tables in this section are based on data collected in 1982 and 1983 from a 100 percent sample of households in the communities. Research conducted on a smaller sample of households (a 50 percent random sample) in the same communities in 1989 and 1990 revealed notable modifications in land tenure. Between the two research periods the two largest landowners in Oroquina increased the size of their landholdings. By 1989 the largest landowner had purchased additional land increasing his total holdings to 110 hectares. The second largest landholder had purchased an additional 10 hectares of land in 1987 for a total of 35 hectares. On the other hand, in 1982/1983, 6 percent of households in both communities were landless (i.e., neither rented nor owned land). The study done in 1989/1990 revealed that by that time the percentage of landless households had increased to 12 percent in Oroquina and to 10 percent in San Esteban. During the interim the amount of land available for rental declined (due, in part, to the changing agricultural strategies of the larger landowners) and rental costs escalated (from US$10.00 to US$20.00 per *manzana* to US$25.00 to US$35.00 per *manzana*. The significance of these findings in terms of augmented concentration of landholdings and increased socioeconomic differentiation will be discussed below and in the following chapter.

12. The sharecropping system, appropriately called *al tercio*, was the same in both communities; a household (male) was given the use of a parcel of land by a larger landowner usually for only one to three cropping cycles. It was the tenant's responsibility to furnish the seed. In exchange, one third of the harvest was returned to the landowner.

13. The micro-level results presented in this section were determined from digitized maps of the research areas which were incorporated into the Geographical Information System (GIS) and the Agronomic Information System (AIS) developed by the Comprehensive Resource Inventory Evaluation System Project at Michigan State University. See Stonich (n.d.) for a description of the methodology employed.

14. Soil analyses were conducted and interpreted in Honduras by staff extensionists at the Honduran Ministry of Natural Resources and in the United States by Dr. Grant Thomas of the University of Kentucky. Most likely the high degree of similarity is because the soils tested were from the same parent materials. Soils from both communities were characterized as having low levels of nitrogen and phosphorous coupled with high levels of calcium, magnesium, iron and manganese. The soils were further evaluated as being relatively fertile but needing applications of nitrogen and phosphorous under conditions of intensive cultivation. The only minor difference were slightly higher levels of phosphorous in the soils at the higher elevations of San Esteban.

15. This cropping system is very similar to that found in the foothill communities of Pespire studied in 1981 and 1982 and reported by B. DeWalt and K. DeWalt (1982) and DeWalt and Stonich (1985).

16. The predominant pasture sown was *zacate jaraguá*, an African grass introduced throughout the Pacific watershed of Central America in the post–World War II period.

17. Without the help of outside agencies these women organized and scheduled the various tasks involved in their cooperative business; e.g., while a few women would be responsible for taking produce to Choluteca other women in the community would assume their responsibilities for childrearing, food preparation, and so forth.

6

Strategies for Survival: The Dynamics of Rural Impoverishment and Environmental Destruction

We (women) are all alone here.

—Southern Honduran woman, 1982

I haven't seen my children in more than six months. They're with my mother in Choluteca. I heard there was work in the new factories in the north but there weren't enough jobs and I didn't have money to get back.

—Honduran woman, 1990

I think the land in the south is good. The main problem is that during the rainy season there is hardly any rain. The rainy seasons there are no longer like the ones we used to have. We're beginning to notice a similar pattern here (la Mosquitia) now. We are already running out of farmland and it will get worse as more people come. We are beginning to notice that our harvests are smaller. Many are noticing that we have less food. If we carry on like this we could end up in a worse predicament than in Choluteca

—Honduran peasant, 1990

Mary and I reached Azacualpa at dusk.[1] It had taken more than three hours to drive the 45 kilometers south from Juticalpa, the largest town in the Department of Olancho—gateway to *la Mosquitia*. It was hot and humid, and we were exhausted. My companion recently had been named director of one of the many nongovernmental organizations (NGOs) operating in Honduras and wanted to see the country. She had worked in Guatemala and El Salvador, as well as in various parts of Africa, but never before in Honduras. Because her job included administering various agricultural and nutritional projects in the south, we began her tour there. That visit had led us to Olancho. Before her arrival she attended briefings and read much of the available information on the region, and believed most of it—that the southerners had given up; that they no longer worked the land and were sitting around waiting for food handouts by USAID, the churches, and the NGOs; and that most southern women were abused by their mates and took little control of their lives. Without diminishing the critical problems that existed in the south, I wanted her to comprehend

123

the ingenuity and courage of southern families in their struggles to make ends meet while being impeded by elites and subverted by the state.

Azacualpa still looks like a frontier town, and until a few years ago it literally was at the end of the road. But no longer. Now it is a juncture where the roads from Juticalpa and Danli converge, and a jumping off point for new settlers from the south. By 1990, many buses per day conveyed new settlers into the region while also acting as the first step in returning cyclical migrants to their homes. A narrow, unpaved, but well traveled road that parallels the *Rio Patuca*, now advances from Azacualpa toward *la Mosquitia*. Between 1974 and 1988, the population of the Department of Olancho grew 86 percent with most of the increase occurring in those municipalities that border *la Mosquitia*; Catacamas (117 percent), Dulce Nombre de Culmi (110 percent), and Juticalpa (94 percent).[2] The towns established by newcomers bear witness to their expectations—*Nuevo Palestino, La Libertad, El Pueblo Nuevo*.

More than half of the men and women we spoke to that evening were from the south—most from the department of Choluteca. We had hoped to drive on further the next morning but were forced to turn back. Gasoline had been expensive and in short supply in Tegucigalpa, and we were not surprised to find that there was no gasoline available in Juticalpa nor in Azacualpa. At the bus stop on the way out of town, we paused to give rides to whomever could fit in the truck. Juan Carlos rode in the cab with us and through him, Mary learned more about the circumstances of southern Honduran families than she could have hoped to learn from reading any reports. Juan Carlos had farmed and worked as an agricultural laborer in El Paraiso and Olancho since the early 1970s. He owned a 3.5 hectare parcel of land in Olancho which he farmed jointly with his 20 year old son, Ernesto. Father and son were traveling with five, 200 pound sacks of corn to *La Garita*, a small village approximately 25 kilometers north of the city of Choluteca where Juan Carlos' *compañera*, Carmen, lived with three of their children. When I asked if he always carried food grains such as corn to Choluteca he replied, "Yes, if we can't grow anything there we can grow it here. Last year we planted corn in *La Garita* but the *langostas* (an array of moth caterpillar larvae) got it. But because I managed to harvest something here we had food and I didn't worry. This year we didn't get beans here (Olancho) but we got them there (Choluteca)."

Juan Carlos also owned 1.4 hectares of land in Choluteca which was being cultivated by his younger sons (aged 12 and 15), with intermittent help from his wife, while he was away. The family owned no cattle but they did own a milk cow that was cared for by his sons. Carmen and their younger daughter maintained a house garden, raised a few pigs and chickens, and made *rosquillas, rosquetas*, and *quesadillas* (pastries made from corn, sorghum, and cheese) for sale. She sold these, as well as milk, cheese, and vegetables, in *La Garita* and in neighboring villages. Carmen also earned income (monetized and nonmonetized) by renting the use of her large outdoor earthen oven to other women in the community.

Because of the expense and the time involved, Juan Carlos had not been to *La Garita* for three months and he wanted to get home very badly. However, first, he had to visit his older daughter (Lucinda), her husband, and their two small children who lived in one of the *colonias* outside Tegucigalpa. His plan was to travel to Tegucigalpa, drop off two sacks of corn at his daughter's house, and then proceed to *La Garita* by bus. His daughter would use some of the corn to make *tamales* which she sold in the central square near the cathedral. Occasionally, Carmen would come into town with pastries, cheese, or vegetables from her garden, and together mother and daughter would sell the commodities. A portion of the cash proceeds from these sales were given to Juan Carlos and Carmen even if they had not provided part of the ingredients from which the foods were made.

Juan Carlos and Ernesto rode with us to Tegucigalpa where we dropped them off at his daughter's house which was in one of the well ensconced squatter communities (*pueblos jóvenes*) north of town. Despite her relief at seeing her father and brother, Lucinda was deeply troubled about the deteriorating economic situation in which she and her family found themselves. The recent devaluation of the Honduran Lempira, the concomitant inflation, and a decline in the number of construction jobs, had left the family with a diminished income. The surge in construction activity in Tegucigalpa which had taken place during the previous few years had ended abruptly. Her husband was a carpenter and had been able to find fairly regular employment at the minimum wage until recently. Now faced with an erratic and insufficient income she found it more and more difficult to buy the ingredients to produce the commodities that she sold. In any case, as she said, most people could no longer afford to buy *tamales*.

Exacerbating the financial picture was the deteriorating health of her family. Tegucigalpa's chronic water shortage had intensified and she had heard that even the Maya Hotel (the most expensive hotel in the city) didn't have water in their bathrooms. Lucinda and her family didn't have a bathroom, and had to purchase water from a truck that came into the barrio at increasingly infrequent intervals. She was convinced that the water they bought was seriously contaminated because all four of the family had been having symptoms of amoebic dysentery for weeks. She felt as if she had nowhere to turn. The clinics run by the Ministry of Health had no medicine and she did not have enough money to take the children to a private physician nor to buy drugs from a pharmacy. She had decided to travel to the North Coast (Puerto Cortes or San Pedro Sula) to see if she could get a job in one of the export processing industries in one of the free-trade zones. While looking for a job she could stay with some cousins from *La Garita* who had migrated to San Pedro Sula five years earlier. Her husband could at times still find work as a carpenter in Tegucigalpa so he would stay. Her major problem was childcare. It was decided quickly that the little ones would accompany Juan Carlos and Ernesto to *La Garita* where they would stay for an indeterminate amount of time until their mother or father came for them. In the meantime, Lucinda and her husband would send what financial help they could to her parents.

Juan Carlos' longer term intentions were to stay in the south until the next planting season, at which time he would return to Olancho. While he was home he would take care of any household maintenance that had to be done—such as making and installing new roof tiles—and help in the upcoming sorghum harvest. Any extra time would be used to make saddles for two customers in Olancho who had given him deposits. Ernesto would help his father transport the grain, visit with his mother and family for a few days, then return to Olancho where he and his *compañera*, Julia, were awaiting the birth of their first child in a few months. Before his father returned Ernesto would begin preparing the family's land for planting. During the interim he would look for additional land with which to expand the family's holdings. Juan Carlos and Ernesto decided that if additional land could be found, 15 year old Roberto would accompany his father to Olancho in the early spring to help clear the land.

The household economy of the family of Juan Carlos and Carmen illustrates many of the essential characteristics found in the economic survival strategies of southern Honduran families, especially those with very limited access to land. Foremost, is the considerable diversification in the sources of monetized and nonmonetized income: subsistence activities; the production and sale of agricultural and nonagricultural commodities; wage labor; migration incomes and remittances; and contractual relationships that result in rental income. Also significant are the myriad economic activities of women and the importance of child and adolescent labor. Apparent, as well, is the contradiction in the preference for neolocal residence and the concomitant economic necessity for extended kin ties. Articulating these diverse components is a flexible household organization—the elastic household—that altered in terms of membership and production strategies depending on a number of factors affecting the adequacy of agricultural production and the availability of wage labor and other cash alternatives.

Diversified Income-generating Activities

The response of peasants to declining real income in contexts of economic scarcity historically has been to intensify existing forms of production (Boserup 1965). In the context of expanding capitalist agriculture and heightened articulation into larger economies, southern Honduran peasants found it essential to combine such intensification (in agriculture) with economic diversification—mixing subsistence production with a complex array of off-farm economic activities. This chapter describes the diversification and organization of the household economy and the associated modifications in regional agricultural systems, and raises critical questions regarding the long-term social and ecological consequences of these linked strategies.

Despite some variation between communities, the array of income generating activities included in the household economies of families in San Esteban and Oroquina were similar, as shown in Figure 6.1 and Table 6.1.[3] A large majority of households cultivated field crops and raised pigs and chickens while about half the families had house gardens. Forty-two percent of households participated in at least one nonmonetized labor sharing network although this was much more prevalent in San Esteban (59 percent of households) than in Oroquina (25 percent of households).

1. Subsistence (domestic) activities done outside market relations which resulted in directly consumable goods including; household maintenance, cultivation of food crops in *milpas* and house gardens, and raising animals.

2. Labor activities that led to the sale of commodities on the market (petty commodity production and commerce); sale of grains, fruits, vegetables, prepared food, animals, animal products and crafts.

3. Wage labor/capital relationships with remuneration in wages or in kind; agricultural and other types of labor in the community, cyclical and long term migration to other areas.

4. Transfer payments that usually took place without immediate reciprocal exchange of labor or commodities; subsidies, gifts, income derived from family members living outside the co-residential unit. These included regular payments (money, goods, services) for maintenance of children and parents, and more irregular payments (money and food sharing) during periods of food shortage and for occasional transport to health care facilities, for school books, fees, clothing, etc.

5. Contractual relationships over the use of land, animals, or equipments that led to rental income; sharecropping, loaning of land, use of mules for transport of crops; equipment for spraying herbicides and insecticides; use of ovens and sugar cane presses.

FIGURE 6.1 Summary of Household Economic Alternatives That Were Pooled into Collective Household Funds. Source: Compiled from survey data.

Petty commodity production and commerce were also common. Sixty-three percent of households reported selling some portion of their domestic harvest of corn, sorghum, and/or beans. Although larger producers were able to sell their products at relatively advantageous prices in regional and national markets, for smallholders, most sales were in-kind debt payments to itinerant, usurious grain merchants (*coyotes*) who often held liens on their crops from debts accrued during previous periods of drought and poor harvests.[4] The percentage of households that sold animals and animal products increased to 60 percent during the period of intensifying drought in 1982 and 1983 as families struggled to augment income. In 1990, however, only 20 percent of the households sampled reported selling any animals or animal products during the previous six month period. In response to questions regarding why more animals were not sold during that time, most householders maintained that they no longer had any animals to sell. The 1980s also appear to have brought

TABLE 6.1 Summary of Household Economic Alternatives: San Esteban and Oroquina

Economic Options[a]	San Esteban n=81	Oroquina n=82	Total n=163
Subsistence Activities:			
Field crops (milpa)	81	76	79
Animals (1982):			
Cattle	19	19	19
Pigs/hogs	63	58	61
Poultry	54	66	62
House garden	41	56	48
Labor sharing	59	25	42
Commodities Production and Commerce:			
Grain (corn, sorghum, beans)	64	62	63
Fruits and vegetables	28	9	19
Animals (1982)	20	48	34
Animals (1983)	47	72	60
Prepared foods	40	49	45
Non-agricultural commodities (e.g.,small stores, pottery)	9	21	15
Wage Labor:			
Male householders	66	68	67
(% from agricultural labor)	(97)	(85)	(91)
Female householders	18	32	26
% of total number of wage earners to household size	57.2	55.7	56.5
Migration			
Male householders	70	83	76
(% who migrated)			
% to rural areas	96	78	86
% to urban areas	4	22	14
Female householders	39	33	36
(% who migrated)			
% to rural areas	40	4	24
% to urban areas	60	96	76
Remittances from Migrant Children:			
% migrant children	31	48	39
Males who send remittances	71	41	56
Females who send remittances	69	66	67
Rental Income:	4	15	9

[a] Percentage of households engaged in various activities.

Source: Computed from survey data.

about a significant decline in the percentage of householders (mostly women) who sold prepared foods to earn cash: from 45 percent of all households in 1982/83 to 25 percent in 1990. As in the example of Lucinda given above, because of rising prices and dwindling incomes, women found it increasingly difficult to buy necessary ingredients. Smaller declines in the percentage of households selling fruits and vegetables (from 19 percent to 16 percent) and nonagricultural commodities (from 15 percent to 12 percent) also occurred by the end of the decade.

In 1982 and 1983, 67 percent of male householders and 26 percent of female householders reported earning income through wage labor during the previous six month period. By 1990 the percentage of male and female householders who reported that they had earned income from wage labor during the last six months had fallen to 47 percent and 8 percent respectively with the mean amount earned approximately US$50.40 (range US$6.00 to US$240.00). Only 8 percent of the male householders interviewed in 1990 reported earning any income through wage work during the previous month. When asked why they did not work more, the majority of men and women (61 percent) responded either that they couldn't find work or that there was no work. A smaller percentage (29 percent) answered that they could not work because of poor health.

Cyclical migration has been an important economic option for most male householders, for more than a third of female householders, and for a growing number of their children. For male householders, two types of migration predominated: the most common was seasonal migration to rural areas that lasted from a few weeks to a few months and was dependent on the availability of agricultural work, and more intermittent but more lengthy migrations (generally lasting from three to ten years) to urban areas and to the North Coast to participate in agricultural and non-agricultural wage work. In contrast to male householders, the majority of women migrated to urban rather than to rural areas, with Tegucigalpa being the most frequent destination. Virtually all women (except the daughters of the group of "rich" peasants who generally left the rural communities to attend school) worked as cooks or domestic servants regardless of where they went. Migration incomes continue to make an important contribution to rural households. Thirty-nine percent of the adolescent and adult children of current householders had left their home communities. Reversing the pattern of their parents' generation more young women than men are emigrating from their home villages. In addition to working as domestic servants, these young women increasingly are seeking work in the export processing zones located far from their families on the north coast. Of the children of current householders, 67 percent of the young women and 56 percent of the young men remitted some cash income to their parents for pooling into common household funds (Stonich 1991a).

Rental income through contractual relationships over the use of land, animals, and equipment was much less widespread. Most of the 9 percent of households who received such income were wealthier members of the communities who had access to sufficient land and animals and who were wealthy enough to own equipment such as pesticide sprayers, outdoor ovens, and sugar cane presses.

Several studies conducted in the south since the mid 1970s, indicate the extent to which household incomes are dependent on off-farm cash earnings. Based on his work in the eastern portion of the southern highlands near the Nicaraguan border during the late 1970s, Boyer concludes that the monetized portion of the total incomes of small holders ranged from approximately 35 percent to 63 percent; that monetized costs extended from 45 percent to 63 percent; and that, in general, monetized costs and incomes were lower in more "isolated" communities than in communities located closer to municipal centers (Boyer 1987). Table 6.2 summarizes the results of a survey conducted among several hundred households in the Choluteca watershed in 1981. It shows that off-farm income contributed from 27 percent to 30.4 percent of total income for households with access to less than five hectares of land. It is significant to note that only farms of more than 20 hectares generated sufficient income to meet minimal subsistence needs as estimated by INCAP. In other words almost all farm families could be classified as living in conditions of absolute poverty. My own, more recent analyses, of household budgets compiled in 1982 and 1983 from households in the adjacent Nacaome watershed indicates that off-farm income contributed from 40 percent to 60 percent of total incomes for households with access to less than five hectares of land (Stonich 1991a).

Analyses of household budgets compiled from 1982 to 1983 from ten households in San Esteban and Oroquina show that the mean percentage of monetized income to total household income was 53 percent (range 31 percent to 87 percent) and of monetized costs 58 percent (range 19 percent to 65 percent).[5] It is, therefore, not surprising that although 80 percent to 90 percent of male householders surveyed considered themselves farmers, virtually all of them had at one time or another earned money through agricultural wage labor; 50 percent reported that they had secondary occupations (e.g., mason, carpenter, and blacksmith); and 76 percent indicated that they had migrated to find work at some time in the past. Women householders also contribute significantly to monetized household income. As shown in Table 6.3, 69 percent of rural women householders took part in some community based enterprise to earn cash—engaging in such diverse activities as small-scale sales of agricultural and nonagricultural based commodities, larger scale sales from small general stores, the performance of a variety of local services, and making crafts. In addition, as discussed above, 36 percent of women householders had left the community temporarily to engage in off-farm income generating activities. Analysis of the information in the household budgets compiled during that same period indicates that off-farm incomes contributed from 20 percent to 60 percent of total cash income for the households studied. The majority of such incomes were from wage labor and remittances from family members who had temporarily emigrated in search of work.

During the 1982–1983 and again during 1989–1990 study periods more than half the sampled households reported at least one family member absent from

TABLE 6.2 Agrarian Structure and Average Farm Income in the Choluteca Watershed: 1981

	Farm Size (hectares)				
	0–2	2–5	5–20	20–50	>50
Percent of farms	---68.2---		21.6	6.8	3.4
Percent of farm area	---8.9---		14.3	13.4	63.5
Mean farm size[a]	1.2	3.1	8.4	29.1	271.9
Sources of income (%)					
Farm income					
Crops	57.9	46.4	52.7	31.2	—
Livestock	11.7	26.7	33.7	59.3	—
Off-farm income	30.4	27.0	13.6	9.5	—
Net income (US$, 1985)					
Per household	366.31	648.25	865.49	1,212.96	—
Per capita	63.16	115.76	144.25	178.38	—

[a] in hectares.

Source: Computed from data cited in García et al. 1988.

the community and engaged in temporary off-farm wage labor. As disclosed in Table 6.4, the major source of household income (including both monetized and nonmonetized sources) reported by female householders in 1990 were wage labor (54 percent of households), subsistence production (20 percent of households), remittances (20 percent of households), and petty commodity production and sale (6 percent). Despite the high degree to which households depended on off-farm income, approximately 80 percent of rural households continued to plant corn, sorghum, and beans.[6]

The Organization of Household Labor

Although the nuclear family is the fundamental residential unit, usually it is not adequate in terms of labor and other economic resources. Family members rely on social relations that reach beyond the rural co-residential unit and articulate them to other residential units often located outside the rural community. Nevertheless, the decisions that rural families make about how to allocate the labor and resources available to them are vital in determining the overall organization of household production.

The Allocation of Household Tasks

The results of a task inventory administered to women and men householders, revealed that very few household tasks were not performed by members of both genders at some time—suggesting that the sexual division of labor is not absolute but has some flexibility.[7] Respondents were asked who usually per-

TABLE 6.3 Types of Income-Generating Activities Reported by Women

Activity	Number of Women[a]	% of Women
Small Scale Sales from Agricultural Production:		
Make and Sell Food	67	44.7
Sell Eggs	42	28.0
Sell Fruits and Vegetables	28	18.7
Raise and Sell Animals	23	15.3
Sell Milk and Cheese	15	10.0
Large Scale Sales:		
Pulperias and Truchas	9	6.0
Services:		
Wash Clothes	27	18.0
Iron Clothes	7	4.7
Seamstress	7	4.7
Injectionist	6	4.0
Practical Nurse	3	2.0
Craft Production and Sale:		
Pottery	3	2.0
Flowers/mats	3	2.0
Other:		
Sell Used Clothes	3	2.0
Raffles	3	2.0
Sell Gas	2	1.0
Teacher	1	0.5
Number and % of women who engage in at least one activity to earn cash	104	69.0

[a] n = 150.

Source: Computed from survey data.

TABLE 6.4 Major Sources of Household Income (monetized and non-monetized) as Reported by Women Householders: 1990

Economic Activity	Number of cases[a]	% of cases
Wage labor	43	54
(agricultural labor)	(40)	(50)
(non-agricultural labor)	(3)	(4)
Subsistence (domestic) production	16	20
Remittances	16	20
Production/sale of commodities	5	6
Total	80	100

[a] n = 80.

Source: Computed from survey data.

formed the task—men or women, adults or children. Of the 22 tasks listed, three were said to be performed exclusively by women and girls (meal preparation, laundry, and childcare) and one exclusively by men and boys (threshing sorghum).

In addition to preparing meals, women have the primary responsibility for the processing of food for domestic consumption; especially grinding and rinsing corn and sorghum. Women are also responsible for daily household maintenance activities such as doing the laundry and cleaning the house. Both men and women care for animals. Pigs and chickens are primarily the responsibility of women who also tend to control any monetary income which results from the sale of these animals or their products. Males are charged with the care of cattle, milk cows, horses, mules, and goats, although women tend to process by-products (milk and cheese) for home consumption and sale. The results of the task inventory indicate that planting, weeding, and harvesting of field crops are the responsibility of males. Women are more likely to participate in threshing and winnowing corn rather than sorghum.[8] Although women's participation in field crop production is low, in general, women's role in the maintenance of house gardens is high. While hauling water is predominantly the work of women and girls, the job of collecting firewood is distributed more evenly among men, boys, women, and girls.

Age is another determinant of intra-household task allocation. Children's work roles are quite extensive and the labor of children is valued. Of all the tasks listed, only one activity was said not to be performed by children. Not surprisingly, the control of family finances is clearly the responsibility of adult family members and is allocated about evenly between men and women. In the allocation of other tasks, however, children figure prominently. Girls' participation is high in the predominantly female tasks such as food preparation, household maintenance, and childcare. Boys, however, perform quite a number of these tasks as well as those that are principally male tasks. In general it is younger boys, particularly those below school age, who perform these "female" tasks. Children below school age tend to spend their time at home with their mothers and both boys and girls carry out tedious and time-consuming domestic chores such as hauling water and grinding grain for the daily supply of tortillas. By age twelve, both boys and girls are thought able to assume full economic responsibilities and perform adult agricultural labor and domestic tasks. They may also migrate to seek agricultural and nonagricultural wage labor or, as in the case of Juan Carlos' middle son, emigrate to frontier areas in order to help expand the family's holdings. The time at which a child reaches adolescence and decisions are made about how to best utilize his or her labor is critical to family survival.

The value of adolescent labor is demonstrated by the plight of young families in the first years of marriage during which they may face their greatest economic challenges. With small children and without the ability to summon adolescent labor, young families are faced simultaneously with high levels of consumption and little access to labor. It is at this stage in the domestic cycle, during which families are supporting young children whose contribution to household economic resources are minimal, that the household dependency ratio is likely to be at its highest.[9]

Intra-Household Time Allocation

Figure 6.2 is based on an analysis of household budget data from both communities. It looks at the allocation of the total labor performed by the entire household, by male householders, and by all other family members according to the percentage of the total number of days per year that were spent in each type of major activity.[10] Figure 6.2.a shows that household maintenance tasks (food processing, house cleaning, laundry, childcare, gathering fuelwood and water, etc.) constituted the majority (57 percent) of the total labor time expended by the households. Next in rank was basic grain production (16 percent) followed by the production and sale of agricultural and nonagricultural commodities (14 percent), wage labor (9 percent), and the production of other agricultural crops and animals (4 percent). The annual apportionment of household labor looks quite different, however, when separated into that done by male householders and by the rest of the family. As shown in Figure 6.2.b male householders allotted the highest percentage of their time to the production of basic grains (38 percent), while wage labor and petty commodity manufacture and vending each took up about the same amount of time (the former 23 percent, the latter 21 percent). Male householders spent very little of their time engaged in household maintenance (11 percent), working in house gardens or raising animals (7 percent). On the other hand, 79 percent the time of other family members were allocated to activities which maintained the household (Figure 6.2.c). A much smaller percentage of their time was apportioned to the production and sale of commodities (13 percent), to basic grain production (5 percent), to other agricultural activities (2 percent), and to wage labor (1 percent).

Rural highland women reported arising between 3:00 a.m. and 5:00 a.m.; retiring from 7:00 p.m. to 9:00 p.m.; and working from 8 to 12 hours per day (mean 11.49 ± 1.35). Table 6.5 rank orders the daily household tasks reported by female householders as the most time consuming. Women were asked to name the three household activities that required most of their time beginning with the task that took the most time. Ninety-six percent of women ranked food preparation as either the first, second, or third most time consuming task. This was followed closely by household maintenance, washing, and ironing (78 percent). Less demanding were childcare (29 percent), collecting water (11 percent), animal husbandry and crop production (4 percent) and fuelwood collection (1 percent).

These results support the information presented in Figure 6.3, which shows the time allocated to various tasks by women householders as a percentage of their total workday.[11] Food processing tasks were the most demanding, accounting for almost 50 percent of women's total work time. Generally, most of that time was taken up making tortillas. Household maintenance (cleaning, mending, shopping, washing dishes, taking food to workers in the field, etc.),

A. Total household labor

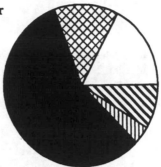

B. Labor of male householder

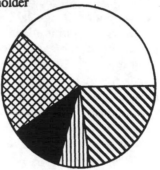

C. Labor of all other household members

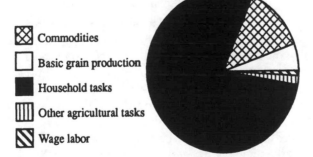

⊠ Commodities

□ Basic grain production

■ Household tasks

▥ Other agricultural tasks

▧ Wage labor

FIGURE 6.2 Time Allocation of Total Household Labor, Labor of Male Householders, and Labor of All Other Household Members as Percent of Annual Work Time. Source: Computed from household budget study.

washing clothes, and ironing also accounted for a significant portion of women's work time (35 percent). Far lesser demands were made by childcare (7 percent), tending animals and crops (3.4 percent), hauling water (2 percent), and collecting fuelwood (.3 percent).

Summary

When looked at altogether, the above time allocation data reinforce the evidence that rural highland families are dependent on monetary income to a great degree. Although male householders spend a great deal of their time (38 percent) engaged in the production of basic grains a significant portion of such grains are sold or used to settle previous debts—for southern families subsistence crops are also cash crops. Moreover, the return on labor invested in basic grain production is low and unpredictable and smallholders are attempting to reduce the costs of production through the enhanced use of chemical inputs which allows them to allocate their labor elsewhere (DeWalt and DeWalt 1982: 31–35). Furthermore, if the two major sources of monetary income for male householders are combined (wage labor and commodity production and vending) the percentage of time allocated to these activities (44 percent) surpasses the percentage of time apportioned to basic grain production. Although a lofty percentage of the labor of the remainder of family members (made up in the majority by the work of women householders) is taken up in household maintenance tasks, 14 percent of the time is used for activities that generate monetary income. Moreover, women householders (such as Carmen) in rural communities are routinely given the responsibility to care for the children of family members who live outside the community in order to enable them to earn wages elsewhere.

Socioeconomic Differentiation and Household Economic Strategies

The organization of households varied around a number of dimensions including the age of the householders and the stage in the family development cycle; family size and structure; the availability of economic resources; and the general pattern in which the particular community was integrated into the region and the nation. As a result, categorizing households for an analysis of various patterns in economic strategies could be based on a number of measures. Land tenure is an especially relevant criteria, however, given the evidence of enhanced land scarcity in the agricultural south. For rural households, the analysis suggests at least six socioeconomically differentiated subgroups and associated economic strategies: a small group of relatively "rich" peasants who are well integrated into the larger economy (approximately 2 percent of all households); a slightly larger group of households (9–10 percent) with access to more than the minimum amount of land thought necessary to sustain rural families in the south (i.e., between 5 to 7 hectares of land depending on quality); the vast majority of households (53 percent) with mar-

TABLE 6.5 Rank of Household Tasks Reported by Women Householders as the Three Most Time-consuming Activities: 1990

Activity	1		2		3		Total	
	N[a]	%	N	%	N	%	N	%
1. Food preparation	45	56	18	23	14	18	77	96
2. Household maintenance	15	19	25	31	22	28	62	78
3. Washing and ironing	13	16	23	29	26	33	62	78
4. Childcare	4	5	10	13	9	11	23	29
5. Collecting water	3	4	2	3	4	5	9	11
6. Animal husbandry	0	0	1	1	2	3	3	4
7. Crop production	0	0	1	1	2	3	3	4
8. Fuelwood collection	0	0	0	0	1	1	1	1

[a]n = 80.

Source: Computed from survey data.

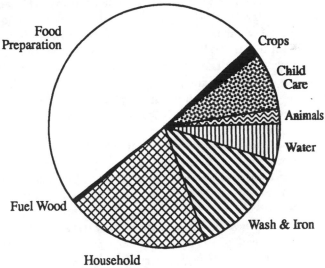

FIGURE 6.3 Time Allocation of Women Householders as Percent of Total Daily Work Time. Source: Computed from time allocation study.

ginal holdings (i.e., between 1 and 5 hectares); a sizable number of near landless households (14 percent) predominated by those who rent at least a portion of the land they farm and who have access to less than 1 hectare of land; and rural households who do not farm (22 percent). This last group can be divided further into one subset made up of women headed households (17 percent) and a second composed of households headed by a conjugal pair (5 percent).

The following case studies based on interviews and surveys completed between 1982 and 1987 illustrate some of the diversity found in the economic survival strategies of households within each group.

Case Study 1: "Rich" Peasant Households

The evolution of a class of "rich" peasants depended on expanding opportunities for realizing capital, particularly through cattle raising and mercantile activities. All three of the households classified into this group were from Oroquina and had augmented their wealth in the post–World War II period through some combination of these emerging opportunities. On the basis of the several measures used to assess household wealth, don Vicente (aged 64) and doña Marta (aged 68) were the most prosperous family in Oroquina.[12] The household was composed of six people; Vicente, Marta, Vicente's 43 year old son Héctor, Héctor's 35 year old wife Lisa, and their two children aged 4 and 7. Lisa and Héctor had two other children who were attending school in Pespire and who returned home approximately once a month.

Don Vicente inherited only five hectares of land from his father, but by the mid 1970s was the largest landowner in Oroquina. He owned 80 hectares of land—94 percent of which was purchased with profits he made raising mules and conducting mule drives throughout the country from the 1930s through the 1950s. He invested the income he earned during these lengthy emigrations in expanded agricultural production—buying cattle as well as land. The size of his cattle herd varied between 50 and 65 head from the early to the mid 1980s and earned approximately US$2,500.00 annually for the family from the sale of cattle, as well as hogs and poultry.

Don Vicente hired local men as agricultural laborers (from 15 to 20 men during the planting season), who were managed by his son, to plant 5–6 hectares of his land in basic grains (corn and sorghum). When needed, he hired additional labor to hand weed, harvest, process, and transport the grains to the regional market. Approximately 85 percent of the corn and 67 percent of the sorghum (either produced on his land or received in payment for rental land) were sold (none to *coyotes*). Most of the remaining corn was consumed by the household while the rest of the sorghum was used to feed animals. The remainder of don Vicente's land was used as pasture or left in fallow. He employed added men to drive his cattle to the Choluteca lowlands during the dry season where he paid to have his herd pastured. The family had numerous fruit trees and a large house garden which were maintained by a woman who was hired full time and who also did the family washing, ironing, and house cleaning. Another woman was employed part-time as a cook.

The family reported eating corn rather than sorghum tortillas for 12 months of the year and grains constituted only 38 percent of their diet—both indications of affluence. Household nutritional status was high: during both the dry and rainy seasons they met more than 190 percent of their household energy and 300 percent of their protein requirements.

Capital accumulation allowed don Vicente to finance community religious and social events as well as economic endeavors, and by the early 1980s he was considered to be the most powerful man in the village—he was don Vicente, the village patriarch. This position, in turn, imposed several social obligations on him, one of the most significant of which was to sell food grains from his stocks to members of the community at below retail prices when their crops failed. He also provided seeds to several families after the protracted drought had forced them to eat their seed stock.

Case Study 2: "Traditional" Farming Households

The case of Pantalión (aged 38) and Francisca (aged 26) comes nearest to representing the "classic" peasant strategy of the southern highlands. According to most estimates of the amount of land necessary to meet household needs the family had enough land (7 hectares) to satisfy their subsistence requirements. Their economic objective was to produce sufficient basic grains and other agricultural products to satisfy nutritional requisites as well as to generate an adequate surplus to meet their limited monetary expenses. Pantalión and Francisca responded vigorously to the high consumption demands made by their six young children; two daughters (aged 11 and 4) and four sons (aged 9, 8, 5, and 2). Francisca inherited two hectares of land in San Esteban from her father to which Pantalión contributed an additional five hectares purchased from his wages harvesting coffee in El Paraiso over a period of ten years. Without outside labor or the use of chemical inputs (they could not afford either) the family cultivated field crops (corn, sorghum, and beans) on two of their seven hectares of land; maintained a large, diverse house garden; and raised a milk cow, a few chickens, and pigs. Although Francisca had never left the community to work she did participate in a variety of community based activities to earn cash: washing and mending clothes, selling milk and dairy products from their family milk cow, and fruits and vegetables from their fruit trees and house garden. Approximately once a month, Pantalión rode his horse to Pespire carrying domestically produced agricultural commodities to sell there or to look for work. He worked not only as an agricultural laborer but also constructed and repaired tile roofs. Less frequently, but on a regular basis, he went to Tegucigalpa, Choluteca, and Nacaome for the same purposes. Pantalión apportioned approximately 60 percent of his labor to produce basic grains in his field; 20 percent engaged in the production and sale of other agricultural commodities; and 10 percent working as a wage laborer. Despite their untiring efforts, Pantalión and Francisca were unable to meet their family's energy requirements both during the dry season (when 82.19 percent of total needs were met) and during the subsequent rainy season (when even less, 75.6 percent of requirements were satisfied). There were several reasons for this. First, the household's high dependency ratio (3.0) taxed the available labor despite the fact that the family engaged in more than 900 days of labor per year. Although the eldest daughter was making an increasingly vital contribution to the household through her work—providing childcare, processing grain and preparing meals, taking charge of the family milk cow, and so forth—her contribution was not sufficient to offset the high consumption demands imposed by the younger children. A second critical factor was the ongoing and worsening drought. Pantalión had lost 75 percent of his previous corn crop to aridity and although the sorghum crop approached a normal one the family had substantial liens on the upcoming sorghum harvest (60 percent of the crop) because of debts accrued during the preceding two years of drought and diminished harvests. The sale of any additional sorghum in the market would provide the family with a source of cash but with little chance of earning enough through wages to purchase food grains afterward. The family dreaded the prospect of being forced to sell even a portion of their remaining harvest. Moreover, they had sold all their pigs and piglets the prior July when their corn crop failed and they no longer had any animals in reserve to sell as a source of cash at a later date. Adding to their dire circumstances was the elimination of El Paraiso as a source of seasonal wage income. Because of the danger posed by enhanced military actions on the Honduran-Nicaraguan border including the activities of marauding bands of Contras within Honduras, 1983 marked the last year that men and women from San Esteban traveled there to participate in the annual coffee harvest. This was a terrible blow to Pantalión as well as to the rest of the two-thirds of the men from San Esteban who counted on migration earnings from harvesting coffee as the principle source of cash for their families.

Case Study 3: Smallholder Marginal Producers

This category encompasses the majority of households, and is distinguished from the previous group by having smaller landholdings (1 to 5 hectares) and enhanced dependence on a diverse set of income generating activities which are integrated through complex and widespread social and economic networks. Although not from San Esteban or Oroquina, the case of Carmen and Juan Carlos presented at the beginning of this chapter exemplifies this strategy.

Case Study 4: Near Landless Households

The activities of Justino (aged 23) and María (aged 19) represent a relatively successful response to conditions of heightened landlessness in the south. It is effective, in large part, because it is built on traditional social patterns that continue to persist to some extent in highland communities. Their household economic strategy is, in reality, shared by two nuclear families—the family of Justino and María and that of Oscar (aged 30) and Pina (aged 26). Oscar who is María's older brother inherited .5 hectares of land from his father and purchased an additional .5 hectares from his mother-in-law at a price considerably below its market value. Justino, too, inherited a .5 hectare parcel. The parents of María and Oscar are dead, and when she married Justino the couple built a small house near her brother. Their propinquity and extensive labor and food sharing activities take on the demeanor of an extended family. In addition to sharing the labor involved in farming their own plots, both men rent additional parcels which they cultivate together with the help of Oscar's 11 year old son. Both men also spend as much time as possible working as paid laborers in the fields of their neighbors.

Pina and Oscar have three other children, all 10 years of age and younger. María and Justino have two children, a toddler and a small baby. Despite the burdens of childcare, Pina's mother, who lives next door, and the two younger women have devised a labor sharing network of their own. While Pina's mother and her oldest daughter care for the little ones, the two young women use the oven that belongs to Pina's mother to make *rosquillas, rosquetas,* savory sweet *quesadillas,* and occasionally *tamales* which they sell in their own and in neighboring communities. Pina's oldest daughter (aged 10) has begun carrying baskets of the prepared foods by bus to Pespire quite regularly (especially during holiday periods and on the days of sports events) where they are sold in the general store located on the highway that connects Pespire to Choluteca and Tegucigalpa. The women reported earning from US$2.00 to US$3.00 a day by selling these snacks—at least the equivalent of the usual day wage for farm laborers.

The two household economies converge not only around economic production and vending operations but also around the sharing of food and other household obligations. In contrast to the case of Pantalión and Francisca who had much more extensive landholdings but who did not participate significantly in any labor, food, or expense sharing networks, the magnitude of interhousehold collaboration between the families of María and Pina suggests the potential effectiveness of extended family ties as a defense against near landlessness. For these two families, the results included impressive income levels for both families especially given their perilous situation: for example, during both the rainy season and the dry season both families comfortably exceeded their household's nutritional requirements.

Case Studies 5, 6, and 7:
Rural Landless and Wage Laborers

The cluster of households without access to land but having both a woman and a man householder represents a relatively small group, but one in which household economic strategies are quite diverse, and household wealth and well-being diverge significantly. The following three brief case studies suggest the heterogeneity of this group.

Cristina and Marco. This household in San Esteban consisted of eight people: Cristina (29 years old), her husband Marco (33), and their six children (aged 12, 10, 7, 5, 3, and 1). They lived in a one-room house made of *estacón* (wooden stakes or sticks) and their material style of life score, at 1, was very low. They had no land although Marco had rented and been loaned land by his cousin in the past. For the previous year, however, the cousin had used the land for his own production. The last time Marco cultivated a *milpa* he planted .35 hectares in corn and sorghum. The corn crop failed because of the drought and the family consumed only 60 percent of the sorghum they produced because of the payment-in-kind they owed the *coyote* who had financed their crop. When possible Marco worked as an agricultural laborer but the family reported that they had not earned any monetary income during the previous month. They had no large animals or fruit trees but they did have a house garden in which they grew vegetables for home consumption. They also owned two hens that produced eggs that were eaten by the family. They ate corn for one month of the year and sorghum during the remaining 11 months: the day they were interviewed, the family's diet consisted of a sorghum/coffee drink (purchased), sorghum tortillas (made with purchased sorghum) and ayote (grown in their garden). The family had a highly grain based diet (86 percent of energy intake) most of which came from sorghum (83 percent). Their household nutritional status was low and declined during the research period: energy intake fell from meeting 57 percent of needs to 43 percent.

In contrast to the previous case of María and Justino, Cristina and Marco have no land of their own and can rely on no one to loan or rent them land on a regular basis; they have a much higher household dependency ratio (3 in contrast to 1); Cristina does not engage in any activities to earn cash, in part because of the demands of her young family; and they do not participate in labor or food sharing networks. In many ways Cristina and Marco are a worst case scenario of the plight of landless families dependent on agriculture.

Liliana and Miguel. Like the previous family this household does not own land but unlike them Liliana (33), Miguel (42), and their four surviving children (ranging in age from 11 to 2), are not dependent on income earned through agriculture. When he was in his early twenties, Miguel left Oroquina and spent two years in Tegucigalpa working as a mason. He then returned to Oroquina and has been earning his living ever since constructing adobe homes there and in neighboring communities. The construction business is seasonal but until 1981 he reported that he was able to support his family with his craft. Since then his employment has been considerably more erratic in a context in which prices have steadily increased. When he works Miguel earns approximately US$40–60 per week about half of which is spent on food for his family. They have no house garden and Liliana does not engage in any monetary income earning activities. By 1982, the family was able to meet only 90 percent of its energy needs and their diet, which they said had been much more varied in the 1970s, had become less diverse and highly dependent on grains (85 percent of calories). In sum, by the early 1980s Liliana and Miguel had already seen a deterioration in their standard of living. In 1986 Liliana and Miguel left Oroquina and migrated to San Pedro Sula in search of work. They have no close relatives in Oroquina and in 1990 no one knew where they were living.

Nicolasa and Adolfo. Unlike either Cristina and Marco or Liliana and Miguel, this family has been able to take economic advantage of the changing circumstances in the south. The family is composed of Adolfo (43), his wife Nicolasa (35), and their five children. The couple's two younger children (4 and 7) reside with them while their three older children (9, 11, and 13) attend school in Pespire and return home on weekends. In the 1970s, Adolfo and his brothers pooled income that they had earned as agricultural laborers and profits from the sale of family owned agricultural land to buy the first in a series of mini-buses and establish a regular bus route from Oroquina to Pespire. The children of one of Oroquina's pioneer families, the brothers had access to enough land so that they were able to sell a portion and still maintain sufficient holdings for family production. In addition Adolfo enjoyed six years of education; only one other male householder from Oroquina (the son of the largest landowner) had attended school for that long. Nicolasa also had gone to school for a relatively long period of time—7 years. While Adolfo drove the bus daily to and from Pespire, Nicolasa ran the family owned general store located at the major village bus stop. The couple's economic pursuits complemented each other and they were able to stock the store at no added transportation cost. The household had many available sources of monetary income: approximately US$30.00 per week earned driving the bus; profits from the sale of food and other commodities (including used clothing from the United States) in their store; and income earned from vending the cattle, hogs, and chickens that they owned (more than US$2,000 per year). Their material style of life score (7) surpassed that of the family of Vicente and Marta (6), the largest landowners in the community. They never ate sorghum tortillas, grains made up only 34 percent of their diet, and they consistently were able to meet more than 200 percent of their family energy and 400 percent of their protein requirements.

Case Study 8: Woman-headed Households

It is impossible to capture in one case study the diversity found in the economic survival strategies of women who are primarily responsible for supplying the needs of their families. Women have many strategies, most involving participation in community based income-generating activities; wage labor; remittances from non-resident family members; integration of older children into the household; and participation in various labor and food sharing networks. Nevertheless the example of Maritza (aged 50) is illustrative and exemplifies the creativity and tenacity that characterize the household strategies of women from throughout the south.

Maritza was born in Oroquina and lived there until 1971 when she and her three small children were deserted by her common law husband. Spurred by the death of her oldest child from diarrhea and the unsuccessful attempt to make ends meet in Oroquina for two years afterwards, she left her remaining children in the care of her mother. She and her younger sister, Angelina, traveled to Tegucigalpa in search of work, where they moved into what was then a new squatter community. Maritza worked as a cook and as a domestic servant while Angelina attended high school and worked part-time. In 1975, Angelina finished school, found a job as a receptionist, and two years later was married. At the death of her mother in 1978, Maritza returned to Oroquina accompanied by Angelina's two small children. There, in the house that had belonged to her mother, Maritza became responsible for the day to day upbringing of her own and Angelina's children. In 1983 she lived with three children—her granddaughter (aged 2), her grandson (aged 4), and Angelina's youngest daughter (15). Maritza has supported herself and her changing family through a combination of activities: working as a milkmaid, seamstress, and cook; producing and vending bread and other prepared foods in Oroquina and in Tegucigalpa; and selling vegetables grown in her house garden. Her son has remained in Oroquina and now has a young family of his own. In addition to sharing food with her son's family, Maritza regularly has charge of his two small children while he and his wife work as wage laborers away from the community.

One cannot understand the economic strategy of Maritza without including the activities of Angelina. Through the years Angelina has provided a vital and regular source of monetary income and other material goods to Maritza, as well as a home in Tegucigalpa for young migrants from Oroquina— including her own and Maritza's children (Stonich 1991a). Angelina and her husband have never returned permanently to Oroquina; however, they have provided guidance, support, and a home in the city, for approximately 20 young rural women and men. In 1983, six young people were living with them. By 1987 eight were resident—four of whom had not been living there in 1983. One of the women who had been living with them in 1983 had moved to San Pedro Sula and another had returned to Oroquina. The San Pedro Sula resident was furnishing a home for young Oroquina migrants in the same way that one had been provided for her in Tegucigalpa. By 1987, the migration network first established by Maritza and Angelina in 1973, included an additional node in San Pedro Sula. There is constant interaction among the nodes: of people traveling in all directions connecting rural communities with both rural and urban wage labor markets; of agricultural products from the rural to the urban areas where they are consumed or enter the market as agricultural commodities and as processed foods; of manufactured goods purchased in urban areas and brought to rural areas; and of regular remittances finding their way to appropriate children, mates, parents, and grandparents.

Social Differentiation, Ecological Change, and Environmental Destruction

For numerous reasons, involving heightened land scarcity, diminished incomes, and mounting populations, the majority of southern families have found it essential to diversify economic strategies while simultaneously intensifying highland agricultural systems and extending their land resource base into frontier regions of tropical humid forest. Monetized incomes have become indispensable and most households participate in a complex array of income generating activities both within and beyond rural villages. Within a relatively short period of time, household members may engage in subsistence production, petty commodity manufacturing and vending, and cyclical migration. In a regional context characterized by the frequent absences of householders and the augmented migration of young women and men, as well as by mounting evidence of widespread environmental destruction, several crucial questions emerge. What are the ecological consequences of social differentiation and of diversified household economic strategies—especially those which require energetic household members to engage in off-farm activities? To what extent have households reached the ecological limits of their efforts to intensify agriculture? Answers to these questions are indispensable in assessing the long-term well being of the people and the natural environment of the south.[13]

The agrarian transformation in southern Honduras included the emergence of destructive agricultural practices on the part of highland farmers having landholdings of all sizes. Data for the region from the national census of agriculture associates a number of such practices to land tenure (Stonich 1992). The percentage of land in cultivation is inversely related to the size of landholdings, while the percentage of land in pasture, the percentage of total cattle owned, and the mean number of cattle owned are positively related to the size of landholdings. It also offers another way to measure agricultural intensification—by the relative use of purchased and/or industrial inputs. It reveals that the larger the farm size the higher the percentage of farms utilizing such inputs. These regional data disclose two broad patterns in agricultural adaptation. Larger farms became more capital intensive, using expanded access to agricultural credit and advanced agricultural technologies in order to secure higher yields from the same land area. At the same time these farms expanded the land-extensive system of cattle ranching. On the other hand as a result of escalating population densities and declining incomes, rural households with diminished holdings were forced to utilize land more intensively by cultivating greater proportions of their land. However, as indicated by Table 6.6, the dichotomy in agricultural practices suggested by the aggregate census data is an oversimplification: community based household surveys revealed that small-scale highland farmers made considerable use of purchased seed and industrial inputs, especially herbicides, in part, as a means of reducing labor costs.

The environmental consequences of differing agricultural practices are extremely important. The interactions among the various agricultural systems that emerged along with each socioeconomically differentiated subgroup (i.e., "rich" farmers, medium farmers, small farmers, and renters) generated mutually reinforcing, destructive effects on the natural environment. Table 6.6 compares a number of potentially destructive agricultural practices to land tenure. It shows that "renters" used their land more intensively than any other group planting 95 percent of the land in annual food crops. It reveals that the most intensive use of land was restricted to small farmers, especially to those having access to less than one hectare of land. At the same time, the proportion of land in pasture and the mean number of cattle owned were significantly related to the size of landholdings.

Within the shifting cultivation systems of the highlands, smallholders are faced with the choice of reducing fallow periods or forgoing fallow cycles altogether; renting land from others if they choose to return their own land to fallow; or substituting wage labor to earn cash to buy food rather than attempt to grow it themselves. Farmers in highland communities exercised all these options. Table 6.6 shows that the average length of the fallow period practiced by

TABLE 6.6 Agricultural Practices by Land Tenure Arrangements, Highland Village Data, Southern Honduras: 1983

Tenancy (ha.)	No.	Land Use and Cattle				Purchased Inputs		
		% of Land Cultivated[a]	% of Land in Pasture	Length Fallow (years)	Mean number Cattle (range)	% Using Purchased Seed[b]	% Using Insecticides[c]	% Using Herbicides[d]
Renters[e]	74	95	—	2.7	0.17 (0—4)	64	7	—
Owners								
<1	23	80	—	2.7	0.22 (0—3)	61	13	30
1–5	87	51	4	3.2	0.22 (0—3)	45	32	28
5–20	15	23	21	3.8	2.5 (0—13)	20	20	7
20–50	2	6	48	5.0	8.0 (7—9)	—	100	—
> 50[f]	1	6	20	6.0	50 (50)	—	100	—

[a] Includes the major food crops—corn, sorghum, and beans.

[b] Fifty-two percent of farmers had purchased seed (corn, sorghum, or beans) in order to resow their fields because drought conditions at the beginning of the agricultural cycle resulted in the loss of their initial planting. Seventy-seven percent of resown maize and 35% of resown sorghum was purchased.

[c] Approximately 2/3 of farmers treated seeds before planting—a process that involved nothing more than putting the seeds into the insecticide mixed with a little water and stirring them around with bare hands. Insecticides were also applied to crops growing in the field. Malathion and Dipteryx were the two most commonly used. They were usually applied with a backpack applicator.

[d] Most farmers weeded twice. If herbicides were used they were applied for the first weeding. The second weeding usually was done with machetes. Few farmers were aware of the herbicide they were using, but the two most common varieties were 2–4–D and Herbisol.

[e] Mean area of rented land = .78 hectares.

[f] Largest landowner rents additional grazing land in lowlands.

Source: Calculations based on survey collected by author. Details on data, methodology, and results available from author.

farmers was directly related to the amount of land they controlled: from a mean of 2.7 years for farmers who had access to less than one hectare of land to 6 years for the largest farmer. During the lifetimes of active farmers, the average length of the fallow period has declined from between 15–20 years to 0–7 years. At the same time, farmers are aware of the potential consequences of shortening the fallow period, and significant differences exist between the ideal and actual practices. While the mean number of years fields were allowed to lie in fallow in San Esteban and Oroquina was 3.5±1.4 (range 0 to 7 years), farmers stated that fields should remain in fallow for at least 8 years (range 4 to 15 years).

Although San Esteban and Oroquina have not been the focus of direct large-scale development projects aimed at combating environmental degradation carried out by either the Honduran government or by the many NGOs operating in the region, residents have had limited access to information about a variety of conservation techniques.[14] Many families own small radios and for several years heard broadcasts promoting governmental and nongovernmental sponsored efforts to stop the practice of burning fields before planting.[15] In addition, for a time during the 1960s and the 1970s, the French Canadian clergy, who vigorously implemented the call of Vatican II for social action, promoted community development and agricultural training through radio schools and teams of peasant agronomists who visited the communities (White 1977). From the late 1970s through the early 1980s, a few agricultural extensionists from various NGOs visited the villages as part of attempts to establish cooperatives and reportedly advocated the building of terraces and rock walls to inhibit soil erosion. Despite the absence of large scale organized efforts, a number of environmental conservation techniques (most intended to combat soil erosion on steep slopes) were implemented within the highland study communities during the 1980s (Table 6.7). The ten most common of these were: leaving trees and saplings in fields before planting (practiced by 46 percent of farming households); employing various slash-and-mulch rather than slash-and-burn methods (43 percent); building check dams (41 percent); sowing field crops across slopes (38 percent); planting more than 50 percent of household field crops on relatively flat parcels rather than on sloped land (20 percent);[16] constructing and maintaining various kinds of terraces (13 percent); and utilizing live barriers (6 percent), ditches (5 percent), animal manure as fertilizer (5 percent), and trash barriers (4 percent). Although for all farms, the mean number of conservation practices was low, 2.2 techniques per farm, the existence of these methods varied considerably in farms belonging to different tenure categories: from a mean number of 0.8 practices on farms of less than one hectare to 6.3 measures on farms of from five to twenty hectares. It is not surprising to find the use of such techniques most widespread on medium size farms: it was this subgroup that invested the most household labor in on-farm agricultural activities. However, despite their "adequate" landholdings and their considerable outlay of agricultural labor, 73 percent of the families in this group failed to meet household energy requirements and collectively satisfied only an average of 92 percent of their calorie needs. Neither should it be startling that the smallest holders practiced the fewest conservation measures:

TABLE 6.7 Agricultural Practices and Techniques Affecting Erosion: According to Percentage of Households in Various Tenure Categories That Utilize Technique or Practice[a]

Practice/Technique	Land Tenure (Total Access to Land)				
	<1 ha. (n=23)	1-5 ha. (n=87)	5-20 ha. (n=15)	>20 ha. (n=3)	Total (n=128)
Trees and saplings left in field before planting	13	44	100	100	46
Do not burn fields	39	37	80	67	43
Use check dams	9	40	87	67	41
Sow across slopes	17	46	20	67	38
>50% of field crops sown on flat land	0	11	93	100	20
Employ terraces	0	2	87	33	13
Utilize live barriers	0	1	27	100	6
Utilize ditches	0	5	13	0	5
Utilize manure	0	0	27	67	5
Utilize trash barriers	0	2	20	0	4
Mean number of conservation practices employed (maximum of 10)	0.8	1.7	6.3	6.0	2.2
Percentage of land in each tenure category according to cadastral maps[b]	1	25	49	24	100

Source: [a] Computed from household survey data. [b] Computed from analysis of digitized cadastral maps of research area.

it is these producers who were most compelled to divide their labor between subsistence activities and a variety of other economic tasks. Under such labor constraints and given the apparent importance of cash in fulfilling household nutritional requirements, it is no wonder that such families chose not to engage in labor intensive conservation measures but rather attempted to reduce labor costs through such means as the use of herbicides and to focus their labor on activities that promised at least short-term increases in income.

Eighty-seven percent of households who farmed less than one hectare of land rented the land to which they had access. Tenants had little motivation to practice conservation measures. The limited amount of land they were allotted was commonly the poorest, the steepest, and the most degraded. Land-owners feared that tenants who used the same parcel for any length of time would be unwilling to leave. As a result tenants normally were allowed to rent a specific plot for a maximum of three years and had little incentive to implement and maintain mechanical soil conservation measures during that time. In addition, after the cultivation period, tenants regularly were required by landowners to sow pasture grass along with the last year's crops, compelling them to become vital links in the forest to pasture conversion process. Moreover rents were high (increasing from US$10–20 per *manzana* in the early 1980s to US$25–35 per *manzana* by 1990 or from one-third to one-half of the harvested crops) and tenants were unable to afford other inputs. Finally, renters were more likely than any other subgroup of farmers to continually burn their parcels, a practice most continued because of the labor demands imposed by the hand removal of thorny undergrowth (Hawkins 1984). In short, the agricultural practices of renters (including the intensive cultivation of the most marginal land; the clearing of all trees and saplings; the repeated burning; and the lack of mechanical soil conservation methods and fertilizers) destroyed soil structure, significantly contributed to erosion, and in total constituted an agricultural system that was extremely destructive to the environment.

To a great extent the detrimental agricultural practices of small, resource-poor predominantly owner farmers (those with access to fewer than 5 hectares of land) were very similar to those of renters. With more secure access to land, however, these farmers were more apt to preserve saplings and useful trees, to construct rock wall terraces, and to undertake other soil conservation measures. Despite the fact that the majority of farmers in this group did not own cattle, crop residues generally were not left to decompose in the field but were collected and used (either consumed domestically or sold) as fodder for other animals during the dry season. It was this group, as well, who were most likely to have attempted to augment their resources by incorporating agricultural plots located in tropical forest zones of Olancho and the *Mosquitia*.

In contrast to landless and land-poor farmers, medium and large farmers maintained longer periods of fallow, were most likely to sustain soil conservation measures, and were least apt to burn their fields. On the other hand, erosion caused by overgrazing, especially during the dry season, and by cattle browsing on saplings and trees in fallow fields, was widespread. It was medium and large landholding farmers who owned most of the cattle and who actively engaged limited-resource farmers in the process of forest-to-agriculture-to-pasture transformation. In the dominant system of pasture management parcels selected for pasture are burned and broadcast sown with seed (*zacate jaraguá*) in the spring of the first season. Beginning in June cattle are allowed to graze freely and most parcels suffer overgrazing that allows the formation of tight masses of tough inedible growth. Tall, inedible flower stalks emerge in November and for the following two months parcels continue to be overgrazed. Cattle scavenge over wide areas establishing trails which are very susceptible to erosion. At the beginning of the following agricultural year the pasture is burned once again in order to permit the grass to reseed and to help eliminate insects (especially ticks). Both overgrazing and burning promote erosion: overgrazing results in the soil being left unprotected and burning damages soil structure.[17]

Population Growth and
Agricultural Intensification

Communities in both the western (Durham 1979) and the eastern (Boyer 1982) portions of the southern highlands were integrated earlier and more fully into the regional and the national economies, than were communities in the central section (the area that includes the municipality of Pespire). Results from earlier studies conducted in those areas, and more recent research in the region around Pespire, indicate that highland agriculture in the south is undergoing a transformation from shifting to permanent cultivation and that this change has been associated with escalating population densities. Table 6.8 shows average changes in population density, in the number of years land was allowed to remain in fallow, and in farming intensity, in the western, eastern, and central portions of the southern highlands. Since 1950, the length of the fallow period has declined precipitously. As the population density of these highland communities increase, there has been a corresponding rise in the intensity of land use (as measured by the intensity of farming). When the population reaches a density of about 100 people per square kilometer, the fallow period disappears and the intensity of farming reaches 100—i.e., the shift from shifting to permanent cultivation is complete. From 1950 through 1988, population densities in the central highlands were lower than those in the western and eastern highlands. Boyer reports that by 1976 more than 50 percent of the households in the communities he studied had eliminated a fallow period entirely and cultivated their fields continuously, and that an additional 27 percent employed only a two year fallow period (Boyer 1982)—something Durham (1979: 144–145) notes for the western highlands at approximately the same time. The farming intensity was the same in 1950 in the communities studied by Boyer as it was in the early 1980s in the central highlands.

These data concerning the relationship between population density and the intensity of land-use correspond closely to those of Binswanger and Pingali (1989: 386). They operationalized Boserup's thesis (1965: 20) that the replacement of forests by bush and grassland is caused by the reduction in fallowing cycles due to increased population densities. According to their data, bush fallow (defined as two or more crops followed by 8–10 years of fallow) and short fallow systems (one to two crops followed by 1 or 2 years of fallow) can take place until a population density of about 64 persons per square kilometer is reached. Above that population density, annual cropping systems are used, and people intensify their use of the land by using fertilizers, animal traction, terracing, and constructing irrigation systems.

Thus, at first glance, data from the south conform to the Boserup hypothesis. The problem, however, is that although farmers in the south are intensifying agricultural production through shortening or eliminating fallow cycles, most are not attempting to maintain the scarce resource (land) by implementing land-conservation measures. In fact, it is those individuals with access to

TABLE 6.8 Relationship Between Population Density, Number of Years of
Fallow, and Farming Intensity*, Southern Highlands: 1950 to 1988

	1950	mid–1970s	mid–1980s	1988
Western & Eastern				
Highlands				
Population				
Density				
(inhabitants/km²)	63	99	110	130
Years of Fallow	3 to 5	0 to 2	0	0
Farming Intensity	38 to 60	60 to 100	100	100
Central Highlands				
Population				
Density				
(inhabitants/km²)	35	54	68	74
Years of Fallow	15 to 20	10 to 15	2 to 6	0 to 3
Farming Intensity	13 to 17	17 to 23	38 to 60	60 to 100

* Farming Intensity = 100 x Number of years of cultivation / (Number of
years of cultivation + number of years of fallow).

Source: Computed from Durham 1979, DGECH 1981, Boyer 1982, Stonich
1986, SECPLAN 1989, and Field Surveys.

the smallest amounts of land who are the least likely to engage in such measures while the few farmers who employ conservation practices in their operations are those who own relatively larger tracts of land.

The consequences of the reduction or elimination of fallowing cycles without concomitant investment in land-conserving technologies are quite apparent in the south. By the late 1970s, both Boyer (1982) and Durham (1979: 140)
report extensive erosion and impoverished soils in the eastern and western
highlands respectively—conditions which spread throughout the region in
the ensuing years. Agricultural yields are declining in many communities, soil
fertility is rapidly depleted, and soil erosion is a substantial problem. The deforestation of hillsides has led to frequent landslides when torrential rains hit
the region. Yet, simultaneously, population increase is not a sufficient causal
explanation for the growing abuse of land, destruction of forests, soil erosion,
or other ecological problems of the region. Rather, inequality in access to land
and the investment patterns of large landowners, neither of which depends
upon population pressure, are at the core of widespread environmental destruction in the region.

Conclusion: Strategies for Survival

Honduran families are not merely victims of economic misfortune but have
conceived innovative strategies to cope with deteriorating circumstances.
These household survival strategies include: (1) diversifying household eco-

nomic strategies; (2) engaging in a wide variety of activities in the expanding informal economic sector; (3) augmenting the participation of women in the work force—both in the informal and the formal sectors; and (4) altering and intensifying their agricultural systems.

Diversifying Household Survival Strategies

As wage labor, cash earning activities, and short and longer term migration became increasingly essential, the number of social and economic ties linking households and communities expanded and became more complex. Agricultural cooperatives, itinerant grain merchants (*coyotes*) acting as middlemen, and community members who travelled outside the community to take advantage of regional and national markets through the sale of their own produce, dairy products, and meat, all functioned as links in growing regional and national networks. The growing number of land poor and landless peasants unable to produce enough for their own needs became integrated into the capitalist agricultural sector through their labor and through the sale of agricultural commodities that they produced. Farmers with medium sized holdings who were able to produce enough for their families had little incentive to produce a surplus of basic grains because of their unfavorable position in the market. Instead, they and the large landholders continued to accumulate land and turned more and more to cattle production. However, for the majority of rural households, who were deficient in land resources and off-farm employment opportunities as well, neither the peasant economy nor wage work alone supplied adequate income. This inadequacy was the product of the environmental constraints encountered in highland areas, the growing population, and the low level of returns to labor and its products. As capitalism spread through the region, proletarianization at the level of the labor process was enhanced, while simultaneously, semiproletarianization arose at the level of the household. The dependence on subsistence agriculture as part of a diversified pattern of economic activities persists. Off-farm activities allow families to improve their standard of living, reduce the risks associated with crop failure, and survive the high-consumption years of childrearing, but they do not provide sufficient income. Families must continue to produce part of their own food.

Participation in multiple economic activities allows the maintenance and reproduction of the household under adverse circumstances. It is not only the diversification of economic activities that is of interest, however, but the various connections that participation in such household strategies institutes between individuals, households, and the larger economy. In this regard, patterns in short and longer term migration are especially valuable in demarcating complex networks that articulate various spheres and levels of social and economic activities beyond the household level: (a) rural communities; (b) urban and rural wage labor markets; (c) petty commodity production/marketing; (d) the "informal" economic sector; (e) and the export processing sector. Moreover, households are most often integrated into multiple spheres. Migration is becoming more essential in the region and migration networks are growing in complexity. At the same time, intra-regional, interregional, and international migrations stimulate the development of capitalism by ensuring the availability of a flexible labor force to meet the seasonal demands of the capitalist agricultural sector.

The process of integrating southern households into new forms of capitalist ventures has been laden with contradictions. These contradictions derive most immediately from the conflict that occurred when new international and national interests entered the region and displaced groups that formerly controlled labor and land. In the long-term, the contradictions are an outcome of the political and economic processes by which dependency was established and of the associated crises of capital accumulation. Honduran families have endured these contradictions in tangible ways. As they commence new cash-earning activities off-farm they find employers averse to paying salaries sufficient for the costs of reproduction; for living and raising a family. Moreover, they find that the prices they receive for their agricultural products continually declining, as subsidized food imports flood the market. If they continue domestic production to supplement these insufficient sources of income they frequently find their available labor resources strained.

The fact that integration into a poorly paid seasonal labor force can be useful for peasants (at least in the short term) explains in part their compliance with an exploitative system. When the ability to meet subsistence needs through agriculture is limited by forces beyond their control, families welcome direct participation in the cash economy as a way to make ends meet. This is germane to understanding why peasant families undertake diversification in household economic strategies. More affluent families with significant resources may assimilate new forms of off-farm activities into their household production systems in order to earn income for potential savings or future reinvestment. In these cases, the nonessential income may be used to buy land or animals, for construction, to finance new ventures in business or transportation, to underwrite important religious or social events, or to serve as a contingency fund. For most families, however, diversification is a response to crises brought about by the destruction of crops by drought or disease, by the less favorable terms of trade associated with the cultivation of food crops, or by the loss of previously available economic activities. These difficulties tend to affect younger families most seriously. Under such circumstances there are few opportunities for saving or for reinvestment.

The Expanding Informal Economy

In Honduras, the informal sector consists of a high percentage of domestic servants and petty vendors, as well as other self-employed workers whose activities range from artisanal production, to food preparation for sale, to making repairs, shining shoes, and performing other services. Although work in this sector normally does not provide a minimum wage, labor protection, or social security benefits, an escalating number of women, men, and children have few other alternatives.

The percentage of rural Honduran women working in the informal sector historically has been much higher than that of men. In 1974 women made up an estimated 59 percent of the informal work force and by 1987 women's participation was estimated to have risen to 68 percent. Very likely this reflects the higher percentage of women who work as petty vendors: trading in home-grown foodstuffs, prepared foods, and locally manufactured goods.

The economically active rural population in the informal sector rose from 9 percent in 1974 to 19 percent in 1987. During the same time period, the percentage in the traditional sector declined from 61 percent to 53 percent and the percentage employed in the formal sector remained approximately the same (30 percent) (Howard-Borjas 1990). Impoverishment certainly would have been higher during the 1980s if it had not been for the growth of the informal economy. The surge in the informal sector may also reflect changes in the policies of international development assistance agencies and the Honduran state which have encouraged the growth of entrepreneurial microenterprises.[18]

The Enhanced Participation of Women in the Work Force

Rural Honduran women are entering the labor force in increasing numbers both in the formal as well as in the informal sector. Between 1974 and 1987 the total economically active rural population grew 51 percent (from approximately 510,000 to 770,000) while the number of women in the labor force soared 278 percent (from 40,000 to 149,000) (Howard-Borjas 1990). The growing number of women in the work force may appear at odds with escalating overall unemployment. Women, however, often are compelled to enter the labor force precisely because of increased unemployment among men, declining real wages, and overall reductions in household income. Several factors contributed to this significant increase in the number of women entering the labor force. Women often are able to find employment even when men cannot because they generally work for lower wages, have more job opportunities in the expanding export processing industries,[19] and because such a high percentage of women work in the informal sector. Whether in the free trade zones located on the north coast or on the shrimp farms and in the shrimp processing plants in the southern lowlands, export manufacturers have preferred to hire women workers because they are cheaper to employ, less likely to raise labor/environmental issues, and have more patience for the tedious assembly line operations.[20]

Alterations and Intensification in Agricultural Systems

The most significant causes of environmental destruction in highland areas of southern Honduras have involved the intensification of agricultural production, poor pasture management, and the expansion of agricultural and pasture lands into steeper, more marginal areas under conditions of rapid population growth. Although pasture is predominant on most of the better lands that are owned by large farmers, poor land management practices resulted in much of this land being overgrazed. On the other hand, small farmers as a result of land fragmentation, declining incomes, and the subsequent reallocation of household labor, increased purchases of industrial inputs, expanded the percentage of their on-farm production sold for cash, and adopted new marketing arrangements. Poor farmers continued to clear steeper and more marginal lands, were less likely to maintain soil conservation measures, persisted in destructive burning, and shortened fallow periods, while being directly brought into the forest to pasture conversion process by larger landholders. The shortening of fallow periods, the continuous burning, the prevailing lack of conservation measures, and the overgrazing that have accompanied forest depletion have led to extensive land degradation (i.e., loss of soil fertility, erosion, and landslides).

Notes

1. This chapter focuses on the household as the primary unit of analysis. The household survival strategies, discussed in the chapter, are linked in important conceptual and methodological ways to the community level analysis presented in Chapter 5. The focus on household economic strategies continues the analysis to more micro-levels and the heterogeneity in household strategies shows the variation in local-level responses to larger scale forces and processes.

Results presented in this chapter are based on ethnographic and survey research conducted between 1981 and 1991. In rural peasant communities the unit employed for the analysis of production and consumption is usually the household often defined as the co-residential unit. For the peasant households in this study, living in a context of advances in capitalist agriculture and increasing incorporation into larger economies, conceptualizing the household as the unit of production is problematic because peasant production does not take place in isolated co-residential units independent of each other, but in households tied together through kinship, cooperation, reciprocal arrangements, tenancy, and wage labor. To a great extent, the maintenance of people living within a co-residential unit is dependent on wage labor, migration, and transfer payments from others living in other co-residential units both within and outside the rural community. for purposes of this research, the "household" was conceptualized not as a co-residential unit, but as a set of relationships between people that imposed sharing obligations on each other. In this context, the household production system (economic survival strategy) was defined as a set of systemic labor relationships among the individuals that comprised the household. What is usually understood as the "extended family" and the "household" overlapped to a great extent, but not exactly, because individuals involved in the reciprocal arrangements were not always family members. For purposes of empirical data collection, however, data were collected from male and female heads of the same co-residential unit. Careful attention was paid to the way and the extent that labor within that unit was integrated with the labor of others living inside and outside the community.

2. These increases far exceeded population growth in Honduras as a whole (65%) and in the south (45%) during the same time period (computed from SECPLAN 1988, 1989).

3. This description is based on the more extensive data collected from 1982 to 1984. Significant differences between data collected during that period and the 1989 and 1990 research period are indicated.

4. According to the 1989 national census of agriculture, 62% of the sorghum, 53% of the beans, and 48% of the corn grown in the south are consumed by producers as human food while 13%, 37% and 43% respectively are sold, and 22%, 0%, and 7% are consumed by animals (SECPLAN 1990).

5. The households from San Esteban and Oroquina included in the budget study diverged substantially in the degree to which they relied on monetized versus nonmonetized incomes and costs. In San Esteban total monetized income averaged US$450.00; the percentage of monetized income to total income was 41 percent; and the percentage of monetized costs was 46 percent. In contrast, in Oroquina total monetized income averaged US$1200.00; the percentage of monetized income to total income was 59 percent; and the percentage of monetized costs was 65 percent. These data provide further evidence of the greater extent to which residents of Oroquina were dependent on cash earnings.

6. These results support a number of other studies of peasant economy in various parts of Latin America that show the extent to which resource poor families with small farms rely heavily on off-farm sources of income. Deere and Wasserstrom (1980) reviewed ten such studies. They found that the average percentage of income derived from off-farm employment ranged from 6 percent in Garcia Rovira, Colombia to 89 percent in Chamula, Mexico and that in five of the ten studies off-farm employment contributed more than 50 percent of total cash income. Further, the proportion of cash income derived from off-farm salaries was greater for low-income families with the smallest land holdings and dropped sharply for higher income families who owned large farms.

7. Twenty-two household tasks were included in the inventory. Beginning with those tasks that were largely the responsibility of women to those that were generally the charge of men, they were; food preparation, laundry, childcare, making and maintaining the family fire, house cleaning, washing corn, washing sorghum, care of chickens and other small animals, collecting water, caring for pigs, grinding corn, managing cash, gathering fuelwood, caring for cattle, caring for goats, caring for horses, preparing land and sowing agricultural fields (*milpas*), harvesting, cultivating and weeding, threshing sorghum.

The same task inventory was administered in 1982/83 and again in 1989/90. The results were essentially the same. A similar inventory was administered throughout Honduras in 1989 as part of a national survey on women. The major differences between my survey results and those of the national survey are: (1) women in the south have significantly more responsibility for managing cash to buy food and other household needs; (2) women in the south have less responsibility for working in field crop production and more in house gardens (Howard Borjas 1990).

8. One reason given for this by male farmers is that sorghum *se pica* (it irritates) when threshed because of its fine chaff. Threshing sorghum takes place during the dry season when the weather is not only dry but also dusty and often windy. Allergic reactions after threshing sorghum include skin and eye irritations. In practice it appeared that the male family member given the responsibility for threshing sorghum is the one who has the least allergic reaction to the activity.

9. For the rural highland sample the Pearson Correlation Coefficient between the age of the female householder and the household dependency ratio was $-.562$ ($p < .0001$) and $-.436$ ($p < .0001$) between the age of the male householder and the dependency ratio.

10. I was unable to adequately separate out the labor of female householders from the labor of the total non-male householders in the family in four out of the 10 budget studies. On the basis of the remaining household budgets, however, I estimate that the labor of the women householders made up from 50–90 percent of the total labor of non-male householder members.

11. Data for this figure were collected through the continuous observation of the female householders included in the household budget study. Realizing the limitations inherent in this method of time allocation (lack of randomness and hence representativeness), it nonetheless represents the way in which rural highland women spend their time. The sample of households used in the budget study were chosen to represent different strata of the rural communities. Each of the ten women householders in the sample was observed four times during one year—twice during the dry season and twice during the rainy season. By that time in the research the range of household activities in which women engaged was fairly well known. The major purposes of the direct observations included acquiring information about the duration and daily flow of different activities. Moreover the results are consistent with the results of a more extensive time allocation study conducted earlier in nearby communities (Fordham et al. 1985).

12. The most important of these measures were access to land, ownership and income from the sale of animals, weekly income, the total amount of monetary compensation earned during the previous six month period, the total amount of money spent per week on purchased food, the scaled score on the material style of life inventory of household goods (possible range of 0 [least wealthy] to 10 [most affluent]), the relative percentages of corn and sorghum in the family diet, and the extent (percentage) to which household energy and protein needs were met.

13. Questions related to the interrelationships between the demands of off-farm labor and ecological change have been addressed by a number of researchers from different disciplines. Deere and Wasserstrom (1981) point out the potential deleterious ecological consequences that come about because the majority of income for small producers in Latin America comes from off-farm labor. Posner and McPherson (1982) contend that procuring major portions of family income from off-farm sources places important restrictions on the availability of labor for farm activities and that any attempts to increase the production of smallholders on steep hillsides must consider the contending demands on their labor that are made by such activities. Collins (1987) calls into question dominant notions of the existence of labor surpluses in peasant communities and attributes ecological decline to the existence of labor scarcity in such communities.

14. Because of the severe ecological disruption, several bilateral assistance agencies and NGOs have, in effect, written off the south and believe that immense, long-term food-aid is the only remaining option for the region. Many more, however, continue to promote their own variation of "development." The largest of these is the Land Use Productivity Enhancement Project (LUPE). With a budget of US$50 million (US$36 million from a grant by USAID the remainder to be provided by the government of Honduras) over the 8-year period 1988 to 1995, LUPE is the largest environmental and natural resource management project currently funded by USAID in Central America (USAID 1989c). The project is implemented through the Ministry of Natural Resources (MNR) and will "incorporate selected NGOs as needed." Stated project goals are "to improve hillside agricultural production and productivity on a sustainable basis, including the management and protection of natural resources on which production depends." The project area covers approximately 18,000 square kilometers and encompasses portions of seven departments—including all of southern Honduras. The so-called "target" group of 50,000 households is among the 250,000 farm families categorized as "Marginal Small Farmers" (<3 ha) and "Commercial Small Farmers" (<5 ha) living in the project area. Families will be reached via a network of 90 extension agencies staffed by at least four extensionists and supported by regional coordinators, technical specialists, and administrators. Up to 80 of these field units will be established and administered by the LUPE project office the remainder administered by NGOs under agreements with the MNR. Activities fall into four project categories: (1) improved cropping systems-

enhancing and diversifying technologies for producing basic grains, fruits, vegetables and tree crops (including nontraditonals); (2) improved animal systems-enhancing animal health, grazing and range practices, and small animal management; (3) post-harvest processing and storage-small silos, pest management, basic handling, processing and packaging at farm and community levels; and (4) facilitated marketing—including assistance to farmers to secure information and credit. Project methods are based on those judged effective in previous more limited USAID projects in the region-primarily the Natural Resource Management Project (NRMP) and the Rural Technologies Project (RTP).

LUPE is ambitious, on paper does not underestimate the severity of the human and environmental problems that exist in the region, and incorporates many worthwhile goals and methods: e.g., it attempts to expand previous effective strategies; it aspires to reduce anti-peasant production and marketing biases; it aims to incorporate NGOs with successful experience; it is directed at poor rural groups that need to increase income; it is based on a mix of technologies that are to be fitted to particular local conditions rather than on a fixed set of technologies to be imposed everywhere; extensionists are to be "generalists" rather than narrowly trained (in the words of the project director); and so forth. However, LUPE has some major shortcomings. Perhaps the most significant is that because of various financial and political delays the project had trouble starting up. In the present global, Central American, and Honduran contexts it is likely that such delays will continue. In addition, the project does not take into account the labor requirements dictated by the multiple income generating activities of the rural poor. Finally, the project dismisses land redistribution as a vital component to increase resources and necessary to enhance rural incomes. In short, LUPE is designed to conform to the larger political-economy which has proved devastating for humans and for the environment.

15. There is considerable diversity among southern highland farmers regarding the occurrence of burning. Because of efforts by the government and other organizations to discourage burning farmers generally view it as destructive, however, in specific circumstances farmers believe that burning is essential to ensure a crop and continue to burn plots. Parcels are most frequently burned before sowing during the first planting season (*primera*) but usually not before the second planting (*postrera*). Burning is also associated with the annual pattern of weed growth. Most weeds and brush germinate or resume growth at the beginning of the rainy season in April and May so that burning at that time eliminates the problem of weeds, brush, and also insects. Because new growth is not as serious a problem during the *postrera* farmers have more flexibility about choosing to burn or to leave the slashed growth as mulch for *postrera* crops. The pattern of not burning during the *postrera* occurs much more frequently on plots which have not been cultivated continuously for several years and where erosion is less severe. Peasants in Pespire and other parts of the south (see Boyer 1991) believe that where soils are depleted not burning results in crop loss. Another traditional limitation on burning concerns cultivating sorghum as a monocrop. By the end of June, weed growth generally progresses sufficiently to allow cutting. At the same time farmers weed they may also broadcast sorghum. As described in chapter 5 sorghum that is grown in this manner commonly is cut and used for animal food. With this system burning does not occur before sowing; however, because southern farmers commonly believe that the sticky sap in the cut stalks inhibits crop growth the following year unless it is burned, after the sorghum has been cut or before the next crop is sown burning does occur.

Although some controversy exists regarding the destructive effects of burning in shifting cultivation systems (Sanchez and Buol 1975) the major problem in the south lies in the lack of sufficient fallow to allow regeneration to a point at which future burns release adequate nutrients. Far less controversy exists about the role of burning in the forest, to agricultural land, to pasture conversion. Owners commonly require that tenants burn the parcels they have been allowed to rent.

16. Land tenure determined who was and who was not able to exercise this practice. It usually was only medium and large landholders who had access to relatively flat parcels.

17. There was a major cause of deforestation that transcended all classes—the harvesting of fuelwood. A total of approximately two million cubic meters of fuelwood are used annually in the south of which 93% is used for cooking fuel in rural villages and where annual consumption exceeds 15 cubic meters per household. Until recently the supply of fuelwood has been abundant: thus the cost to consumers reflected only labor and transportation and did not include the value of the raw material. This situation changed dramatically during the 1980s as supply areas were depleted of their reserves and shortages became more acute. By 1990 even in remote villages people were experiencing difficulties in obtaining needed quantities.

18. This has been true throughout Latin America and the Caribbean. Because of their low cost and their capacity to absorb large amounts of labor, microenterprises have been perceived as having a comparative advantage in a competitive international market. The development of microenterprises has been given increasing support from both the public and private sectors in terms of credit, access to raw materials and foreign exchange, and other privileges previously reserved for the formal sector (Safa 1987).

19. In addition to the efforts of USAID to promote nontraditionals a number of Honduran decrees offer incentives to exporters. The Puerto Cortes Industrial Free Zone, created by Decree No. 356 in 1976 offers exporters duty-free importation of machines and raw materials, no federal, state, or local income or corporate taxes, unrestricted repatriation of profits and capital, and exemption from export controls and duties. The Temporary Import Regime (Decree No. 87 1984) suspends import taxes on raw materials and equipment and eliminates the income tax for ten years for producers who export outside Central America (Paus 1988). In 1986 the incoming Liberal Party president Jose Azcona outlined a plan to promote export-led growth, in part, by expanding export processing zones which have been augmented further by the National Party President Rafael Callejas since his inauguration in 1990. To encourage additional investment, Honduras offers tax-free status *forever* in its industrial parks and free zones and duty-free status on all imported machinery and raw materials, as well as any value added locally. Honduras also promotes *private* free zones which are industrial parks built by developers then marketed to foreign investors free of state control. In mid-1991 five such parks were under construction outside of San Pedro Sula and by 1992 additional parks were planned for Choluteca and San Lorenzo.

20. The benefits for firms employing women in the production of agricultural and industrial exports in Latin America is well researched (e.g., Arizpe and Aranda 1986; Nash 1986; Safa 1986; Deere 1987).

7

Conclusions:
The Political Ecology
of Development

Current human and environmental crises in southern Honduras are the outcome of a long history during which the region became integrated into national and international systems. The political ecology that evolved over centuries, significantly affected the distribution and use of resources, precipitated dynamic contradictions between society and natural resources, and stimulated social and ecological change. These effects can be traced through the interrelations among multiple levels of analysis—from more macro-levels to more micro-levels. While powerful macro-level actors (international, national, and regional) exerted considerable influence, actors at the local level (community, household, and individual) mitigated the impact of those forces in vital ways. Local level efforts, in turn, have profoundly affected the local actors and the natural environment in which they live and prompted change within the region and the nation as a whole.

The articulation of southern Honduras into more extensive systems was quite uneven over time, space, and peoples. Although complete incorporation did not ensue until after World War II, the foundations for contemporary patterns in settlement, social relations, and agriculture were instituted during earlier periods. The Spanish conquest had immediate and devastating consequences on the indigenous peoples of region and by the end of the 16th century the lowland south was virtually depopulated while the highlands became the site of more dense settlement by remnant indigenous populations. The lowlands became the location of cattle ranches controlled by a few large and, much more numerous, medium size landowning Spaniards who eventually became the patriarchs of the regional elites. In the highlands, indigenous people quickly became ladinoized and survived by means of their own agricultural production and by working as laborers in mines or on *haciendas* during a series of subsequent boom and bust cycles in mining and agriculture. When the world market provided advantageous opportunities for new export crops, regional elites used their power to acquire suitable land and often evicted peasant producers from the desired holdings. During succeeding bust cycles, peasants reclaimed the expropriated land and used it for their own subsistence production.

These colonial patterns persisted until the mid-19th century when socioeconomic and demographic alterations began to arise in conjunction with the re-opening of mines in highland areas. Many of the highland localities that currently have the highest population concentrations were the sites of mines started either in colonial times or during this period. However, until 1900, settlement in the south consisted primarily of large and medium sized cattle ranches in the lowlands and scattered peasant homesteads in the highlands. By the 1930s, subsistence production continued to predominate in highland peasant communities but production for exchange began to expand due to increases in the size and the number of settlements (that were brought about by natural population growth and in-migration to the region) and by the proliferation of merchants. Despite these changes, southern Honduras remained for the most part economically disarticulated from the rest of the nation, and in turn, from the world.

More absolute integration of the south came only at mid-century following attempts to stimulate economic growth throughout the nation during the post–World War II period of high prices for agricultural commodities. These initiatives included establishing a regional transportation system, organizing markets, and providing expanded sources of credit. With these measures in place, large landowners who historically had been unable to respond to favorable external economic conditions found it profitable to expand production for the export market. Since mid-century, economic development in southern Honduras has stressed the diversification of export-led agriculture.

Promoted by bilateral and multilateral lending institutions, transnational corporations, and the state, the growth of nontraditional export commodities in southern Honduras during the 1960s and the 1970s, especially cotton and cattle, exacerbated inequalities and social differentiation. The promotion of cattle ranching was especially significant because it stimulated the expansion of capitalized agriculture into highland areas. Although violent local clashes occurred frequently between southern peasants and authorities, the actions of the national government helped diffuse their impact and conflict did not escalate to the same levels as in neighboring El Salvador, Nicaragua, and Guatemala (Booth and Walker 1989). Nevertheless, the export-led development model was unable to create adequate employment due to fluctuations in the world export market, land concentration, mechanization, and the replacement of crops by extensive livestock operations. Even though the intensity of violence in the south never approached that found in adjacent nations, the degree of impoverishment in the south surpassed that of its neighbors. The semiproletarianized peasantry that emerged during that period had access neither to sufficient land nor to employment opportunities to reduce poverty. The ensuing agrarian structure consisted of a complex network of classes and interest groups each possessing various amounts of social power and commanding different proportions of natural resources. The more powerful actors switched from investment in one export commodity to another with little regard for the environmental and social costs and took advantage of the opportunities provided by the newest wave of nontraditional exports.

The promotion of the latest trend of nontraditional exports in southern Honduras, as well as in the rest of Central America, was superimposed on an agrarian structure already in place. Social and ecological problems comparable to those associated with commodities promoted earlier have begun to appear throughout Central America. The contention that small producers would benefit proportionately from the promotion of nontraditional exports is being questioned and apprehension over the environmental consequences related to the production of nontraditional exports is rising (Kreuger 1989; Murray 1991, Rosset 1991, Murray and Hoppin 1992). Such concerns include elevated incidence of pests and diseases because of the introduction and the intensive cultivation of crops; resulting ecological imbalances; crop, water, and air contamination stemming from improper use of pesticides and other chemicals; and the creation of "pesticide treadmills" brought about as producers escalate their use of chemical inputs in response to the increased resistance of pests to chemical strategies (USAID 1989d). The excessive and indiscriminate use of pesticides, a significant problem for more than thirty years in cotton and other export crops has become common in the production of nontraditional export crops throughout the region. This pattern of pesticide dependency and misuse poses a serious economic threat to the long-term development of the nontraditional sector because growing numbers of produce shipments from the region have been confiscated and destroyed upon entry in the U.S. due to the presence of illegal pesticide residues (WHO 1990; Hansen 1990).

Human Agency and Environmental Destruction

The dominant development model has not been able to solve the fundamental crises in southern Honduras. The political-ecology of previous development policy has left the bitter legacy of a spatially marginalized population, a highly vulnerable economy, an extremely unequal distribution of access to resources and income, widespread food insecurity, and rampant environmental degradation. Moreover, the economic crisis of the 1980s and the structural adjustment measures employed to cope with it, have exacerbated these interconnected crises. Although declines in standards of living have been most severe for the poor, the middle class has also suffered from widespread cutbacks in government employment and services. Although powerful forces have constrained potential local level responses, the people of the south have acted in order to temper the effects of external forces and, in turn, human agency is influencing the larger system in significant ways.

Diversification of the household economy, enhanced dependence on the informal sector, and the increased participation of women in the labor force, have centered attention on the household as the place where strategies for survival are conceived, organized, and implemented. In the context of ongoing and worsening crisis, households, such as those of Juan Carlos and Carmen, Justino and María, Maritza and Angelina have become the nuclei for small-scale entrepreneurial activities: preparing food for the market, growing and selling agricultural commodities, selling lottery tickets, producing crafts, sewing, and engaging in the myriad other activities that are part of the informal sector. In addition, families attempt to expand these enterprises by making use of extended family, local, and extra-local social and economic networks. As the case studies have shown, these networks have been important economic links and have enabled households to respond imaginatively during times of hardship. At the same time, international and national investment, as well as government policy, are able to take advantage of households' capacities to rely on the informal economy and on complex social networks to augment insufficient income. Thus, by mitigating the effects of an exploitative system, creative household strategies also encourage continued exploitation and the inequitable distribution of resources which precipitated the various crises to begin with.

Households are also basic to understanding the interrelationships between development and enhanced destruction of the natural environment. In the southern Honduran case, destructive management of resources by smallholders is frequently the result of the increasingly diversified nature of household production systems and in particular of the fact that producers must divide their labor between work in the fields and a variety of other vital economic activities. Under conditions of labor scarcity, farmers may abandon agricultural practices that have proven effective over time, in favor of techniques with lower demands on labor as well as the promise of short-term increases in production. The decisions of the more powerful actors in the region which have resulted in widespread overgrazing, pesticide contamination, and destruction of highland, foothill, lowland, and coastal resources are no more sustainable in the long-term.

The high price of environmental degradation is shrouded by a number of factors including the resourceful survival strategies of peasant households. The long-term consequences of environmental destruction for such marginal peasant populations already suffering from food insecurity promise to be profound, yet these populations are the least able to confront the issues and redirect development strategies which continue to deplete the natural resource base. The crucial importance of peasant farmers to agriculture and to national food security should not be underestimated: it is small and medium farmers who grow the majority of the basic human food grains despite low yields and severe anti-peasant production and marketing biases (Stonich and DeWalt 1989; Stonich 1991c). The progressive depletion of the natural resource base which supplies the majority of corn, sorghum, and beans to Hondurans throughout the nation will cause further escalation in the dependence on food aid and food imports.

As summarized in Figure 7.1, the interconnections among economic development efforts, rural impoverishment, and environmental destruction in southern Honduras are complex. Since mid-century, the region has been distinguished by expanding capitalist social relations that have transformed the agrarian structure and have shaped the ways in which natural resources have been exploited. Those national groups that hold positions of wealth and privilege also tend to have a prominent voice in the area of public policy; many are

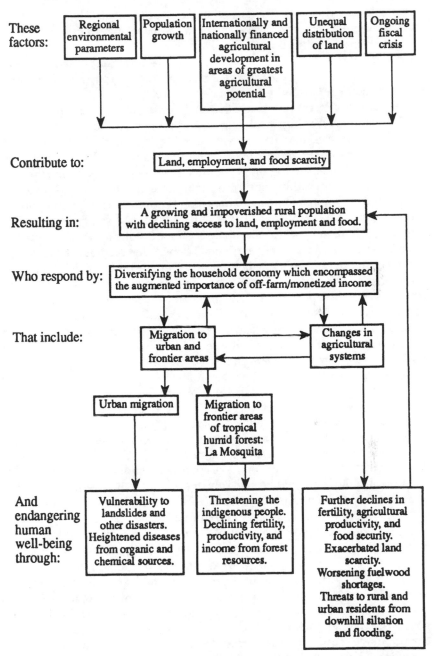

FIGURE 7.1 Interconnections Among Development, Rural Impoverishment, and Environmental Destruction in Southern Honduras. Source: Compiled by author.

important officials in the government and in the military. Consequently, the instruments of the state tend to gratify the requisites of the dominant classes. The effects of environmental destruction are often immediately imperceptible, slow to proceed, complex to measure, and spatially uneven in their distribution. More influential groups normally are in the best position to adjust to these negative effects and also the least likely to experience them on a daily basis—inhibiting the ease with which such powerful groups might be mobilized around issues of environmental destruction.

The results of international and national forces in the region provoked responses on the part of highland farmers and these reactions encompassed agricultural strategies that have accelerated ecological decline. Small-scale, resource-poor farmers often directly cause environmental deterioration because they are under such great pressure to secure a livelihood, that the short-term costs of conservation efforts are prohibitive. It is necessary, however, to understand the systemic interrelationships among different power holding groups before determining the causes, directions, and solutions to environmental problems. It is overly simplistic to single out the agricultural practices of one or another group when ascribing "blame." For example, the techniques of renters and the smallest landholders appear to be the most destructive; yet, given their environmental, economic, and demographic constraints such farmers have few options and often are compelled to act as agents of more powerful, larger landholders in the process of forest-to-pasture conversion. Ironically, as shown by the endeavors of medium size farmers from Pespire, implementing more environmentally sound farming methods does not necessarily lead to income which is sufficient to meet household nutritional needs. In their case, the contradiction between maintaining the natural environment and sustaining the household is striking. In southern Honduras, environmental degradation is tied to the structure of society and intricately linked to problems of land tenure, employment, demography, and poverty.

Although the rapid increase in population growth in the region is a matter for serious concern, population growth per se cannot adequately explain the destructive land-use patterns that have emerged in the south. Agricultural development in the region has been highly uneven not only in terms of the distribution of economic costs and benefits but also in terms of its effects on the spatial distribution of people. Political-economic factors related to the expansion of export-oriented agriculture constrain access to the most fertile lands of the region. The growing population in the highlands with inadequate opportunities to earn a living continuously apportion the area among more and more people, intensify their agricultural production, and expand into even more marginal areas. Growing rural poverty stimulates out-migration from the more densely packed south into other parts of the country, thereby decreasing population pressure within the region while simultaneously augmenting urban populations and escalating pressure on tropical forest areas in the remainder of the country. Within the south, in urban centers throughout Honduras, and in frontier areas open to resettlement, the mounting evidence of ecological and human decline emanating from these various processes magnify the threats to long-term human, economic, and environmental sustainability.

Beyond the Prevailing Development Model

During the 1980s, agriculture emerged as the most dynamic sector of the economy in Central America and the percentage of earnings from the export of natural resource based commodities rose in every Central American country. It may be possible to alter development priorities without abandoning export agriculture and thereby take advantage of the important role played by agriculture while simultaneously decreasing rural impoverishment and environmental degradation (see e.g., ECLAC 1986; Paus 1988). There are sufficient warnings, however, that fundamental changes are urgently needed in the current development strategy. Among the interrelated group of factors that could help affect such changes and promote growth with equity at the local level are: applied political-ecological analysis; improved access to productive resources —land and employment; a development approach that is based on the differentiated rural society and diversified poverty and which recognizes the link between household production and reproduction; and strengthened local organizations.

Applied Political-Ecological Analysis

While it is apparent that development is a political process, practical recommendations often do not go beyond this assertion. An alternative is to treat power with appropriate respect by identifying the various groups and interests involved in the development process; assessing their influence and their relative control of resources; tracing the linkages among them; and prescribing ways to negotiate and mobilize support. The probability of effective interventions would rise if these contending forces were dealt with candidly during the process of policy design and implementation.

Too often discussion surrounding economic policy is conducted at a relatively high level of abstraction. This includes recent debates on how best to deal with Central America's diverse problems. Positions are defended with very little reference to the actual political constraints surrounding policy formation or to the complexity of local societies upon which policies eventually encroach. In the end, however, the ultimate success or failure of policies are determined by these social and political considerations.

Improved Access to Productive Resources

A principal cause of rural poverty is the lack of access to sufficient land and the low productivity of land that is available to most people. Any strategy aimed at reducing rural poverty must augment access to productive resources involving both land and enhanced opportunities for off-farm employment. Although the national agrarian reform program begun in the 1970s is often cited as one of the primary reasons for the relative political stability in Honduras (e.g., Williams 1986; Booth and Walker 1989), the stagnation of the program since the late 1970s contributed to subsequent peasant political radicalization (Ruhl 1984, 1987). The recent passage of the agricultural modernization law already has aroused similar actions. Unfortunately, recent Honduran administrations have been unwilling to bear the immense financial costs of colonizing large tracts of new land, appropriately, and rehabilitating degraded areas for settlement, or to confront the enormous political obstacles to expropriating unused private property from large landowners under the provisions of the agrarian reform law (Ruhl 1987).

Development Based on a Heterogeneous Rural Society

Rural society and rural poverty are highly differentiated among and within communities; accordingly, there cannot be *one* program to combat rural poverty. Programs must be aimed at increasing the income generating activities which comprise the economic strategies of different categories of rural households. Such approaches also must consider the value of women's unpaid domestic labor and the articulation between economic production and reproduction in peasant households.

Landless households and those with access to less than five hectares of land comprise the majority of southern Honduran households—a semiproletarianized group who juggle the labor demands imposed by multiple farm and off-farm economic activities. Merely increasing agricultural productivity is not an adequate solution for households who tend to be tenants rather than owners and who earn most of their income from off-farm sources. Strategies aimed at reducing poverty among this group must incorporate issues of access to land but also those of employment creation and off-farm wages. Such efforts should be focused at the household level, fit the labor demands dictated by multiple income generating activities, and take into account the value of the unpaid domestic work of women and their enhanced participation in the labor force. Participation of women in either the informal or formal market sector enhances household income; is likely to be invested in children's health, nutrition, and education; and hence raises the standard of living for the household (Danes et al. 1987). It is vital that an alternative development approach take into account the articulation of household production and reproduction, improve conditions of human reproduction, and enhance the conditions under which domestic labor is performed through augmented social services and basic infrastructure, more appropriate and environmentally sound technologies to perform domestic tasks, and rectified inequalities in the division of labor by gender. Envisioned technical innovations must be evaluated in relation to the risks they present and the opportunity costs of lost wages if family members are restricted from participating in seasonal labor markets. With viable alternatives such households are less apt to continue the environmentally destructive farming practices discussed above.

Projects aimed directly at increasing agricultural productivity and improving degraded lands are most appropriate for households with medium holdings (i.e., more than five hectares of land). Development efforts should emphasize augmenting production through improved cropping and animal systems (including nontraditional commodities where appropriate), better postharvest processing and storage systems, and enhanced marketing arrangements including information and credit. Programs should also include specific strategies directly aimed at improving degraded lands. Many of the active development projects in the region financed by NGOs, USAID, and other bilateral assistance agencies already focus on this group and embody many of these objectives. But isolated projects in areas such as "sustainable farming," soil conservation, or integrated pest management are not sufficient to remedy the human and environmental disasters in the south.

Community Organizations and the State

Largely in response to the failure of the state to address the needs of local communities, social networks linking the multiple economic activities of households are growing in vitality and complexity (see Stonich, 1991a). Development efforts should promote such local initiatives through which the communities can plan their own futures with dignity, make use of the human and material resources within the community, integrate communities into the region and the nation in more beneficial ways, and potentially generate income and expand social services. The promotion of such community organizations must assure the active membership and leadership roles of women whose escalated participation in the informal and the formal economic sectors constituted the majority of the gains made in those sectors during the 1980s. The expanded and crucial role of women in the labor force and the critical value of women's unpaid domestic work requires that special attention be given to the differential impact of development policies on women and men. Primarily with the help of NGOs, highland and coastal community groups such as CODDEFFAGOLF, already are attempting to achieve some of these goals: greater political and economic leverage for local people by exercising influence over local administrators and asserting claims on government; adapting project activities to local conditions; and managing the natural resource base rationally through education and training.

However, organizations such as CODDEFFAGOLF, do not have the power of the state, and their achievements are determined, to a considerable extent, by their position within the larger political ecology. In the case of CODDEFFAGOLF, despite garnering outside support from the Honduran public, from academics, and from national and international environmental groups, tensions persist between the organization and the Honduran government, the large shrimp farmers, and the USAID funded export diversification program.

People, Poverty, and Environment in Central America

Although the focus of this book has been on Honduras, all Central American governments have a long history of pursuing development policies that support the expansion of agricultural commodities for export. To a great extent the underlying and as well as the more immediate causes of the current crises throughout the region include fundamental structural inequalities, the expropriation of peasant lands, and the capitalization of agricultural labor relations that issued from export promotion. The agricultural export-led growth strategy led to the concentration of land and other forms of wealth in the hands of a small minority and to the progressive impoverishment of much of the rural population. Families with diminished access to land became increasingly dependent on monetized incomes earned off their farms especially on seasonal wage labor on large estates.

Recognition is growing of the systemic links among this model of development in Central America, accompanying high levels of poverty, and intensified environmental degradation. Nevertheless, throughout the region, governments, households, and individuals alike are over-exploiting the natural resources that they control in order to generate income to satisfy immediate needs, thereby increasing pressures on resources at the same time that knowledge of the interactions among these factors expands. The cumulative effects of unrestrained agro-export led development and peasant populations desperate to sustain their families may well devastate much of the natural environment of Central America. The consequences of deforestation, erosion, watershed deterioration, pesticide abuse, and other forms of environmental destruction are becoming conspicuous, but there are few indications that the present development strategy will be altered in any significant way.

There is increasing evidence that the tactic of export diversification, as a means to allay the varied crises of the 1980s, has enhanced or augmented many of the social, economic, and environmental circumstances that were the basic foundations of these crises to begin with. Export diversification toward nontraditional products may be one important element of a new model of development for the region, but it cannot be the focal point in the search for less vulnerable economies. Moreover, reforming economic relations around a division of labor based on the supposed Central American comparative advantage of cheap labor may increase the profits of U.S., Japanese, and Korean corporations but will not necessarily raise the standard of living of Central American people. The continued concession of long-term sustainability in favor of short-term gain and the persistent denial of the disastrous social and environmental effects of the prevailing development model, may induce ecosystems to surpass their limits and stimulate lasting human and ecological crises that are impossible to resolve.

Appendix:
Research Methodology

The information in this book comes from a variety of sources including fieldnotes, surveys, interviews, household budget studies, maps, computerized satellite imagery, aerial photos, and documentary/statistical materials. This appendix describes the overall methodology used in the research. For further details on the integration of data from remote sensing devices and the use of Geographic Information Systems, see Stonich (n.d.) and for the dietary/nutritional analysis, see Stonich (1986, Appendix A).

The major objective of the methodology was to create a systemic framework in which to integrate the relevant factors and levels of analysis involved in the social and ecological processes related to the expansion of capitalist agriculture (Figure A.1). The goal was to integrate a representative sample of micro-level studies (of individuals and of communities) into a hierarchical structure composed of multiple, relevant, and increasingly macro-levels of analysis. An additional objective was to add a historic dimension by positioning the traditional micro-level foci of anthropology—individuals and the community–at the intersection of local and world history. Sampling strategies were designed to reflect the underlying assumption that each level and factor was heterogeneous.

The choice of specific levels of analysis from which data were gathered was based on theoretical considerations but also influenced by the conceptualization and construction of levels of analysis reflected in the data available from national and international sources. For example, little relevant data (social and/or ecological) had been collected by such agencies and/or were available at the levels of the individual householders or particular communities. Rather such agencies provided information at the levels of the municipality (similar to counties in the U.S.), the department (similar to states in the U.S.), and the region ("the South" is treated as a separate region of the country as defined by national and international agencies). As much as possible, both qualitative and quantitative data were collected from such agencies in the forms of written materials (e.g., published and unpublished documents, statistics, maps, census data, and so on) and interviews with administrators and officials. Considerable care was given to the collection of data from such sources both early in the research process and continuously throughout the study and the contacts made during the initial search and interviewing process proved invaluable. Given the macro-level focus of data from such sources, the core of the an-

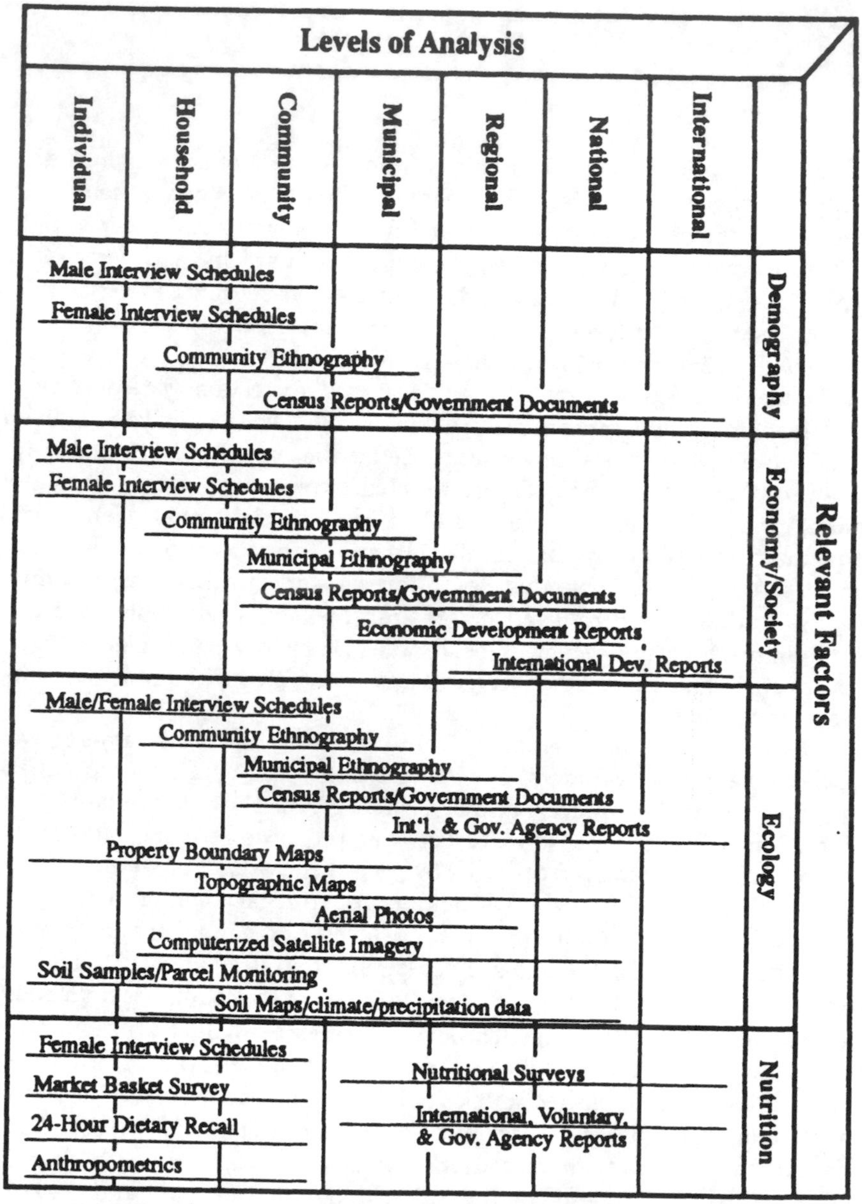

Levels of Analysis

Individual	Household	Community	Municipal	Regional	National	International	Relevant Factors

Demography

- Male Interview Schedules
- Female Interview Schedules
- Community Ethnography
- Census Reports/Government Documents

Economy/Society

- Male Interview Schedules
- Female Interview Schedules
- Community Ethnography
- Municipal Ethnography
- Census Reports/Government Documents
- Economic Development Reports
- International Dev. Reports

Ecology

- Male/Female Interview Schedules
- Community Ethnography
- Municipal Ethnography
- Census Reports/Government Documents
- Int'l. & Gov. Agency Reports
- Property Boundary Maps
- Topographic Maps
- Aerial Photos
- Computerized Satellite Imagery
- Soil Samples/Parcel Monitoring
- Soil Maps/climate/precipitation data

Nutrition

- Female Interview Schedules
- Market Basket Survey
- 24-Hour Dietary Recall
- Anthropometrics
- Nutritional Surveys
- International, Voluntary, & Gov. Agency Reports

FIGURE A.1 Relationship Among Various Sources of Data, Relevant Factors, and Levels of Analysis. Source: Compiled by author.

thropological research was the study of individuals and communities—the traditional loci of anthropological research. The objective was to obtain data from a sample of communities that were representative of the variation in social/environmental interactions that existed in the region. After a preliminary study of written documents, numerous interviews with officials, and a short period of initial field research throughout the south, a sample of nine research communities were chosen to represent the agroecological variation present within the region. Using a combination of ethnographic (including both qualitative and quantitative observations) and survey research methods, primary data were collected from 50 percent random samples of men and women householders in these communities in 1982 and 1983. For the two communities described in detail in this book, the sample was extended to 100 percent of all households. This was done in order to facilitate the more complete understanding of the complex social and economic networks operating within the communities. Limited supplementary ethnographic research was conducted in the communities in 1986 and 1987, followed by another round of survey and ethnographic research in 1989 and 1990.

Interview schedules were administered to women and men householders at least twice—once during the dry season and again during the rainy season in order to obtain information on seasonal variation in economic/agricultural practices, diet, and nutritional status. They also reflected changes due to a continuing and worsening drought, increased militarization, and local responses to the worldwide economic recession that was having a very serious affect throughout Central America. The women's schedule emphasized economic resources available to women (including information on migration and a labor history), agricultural practices, household composition and family history, reproductive history, group memberships, beliefs about health, and use of health care facilities. Data concerned more directly with food use were also gathered. A market basket survey using a list-recall format and 24-hour dietary recall of family meals and for all children under 60 months of age were also administered to women (see below). Anthropometric measures were also taken on all children under 60 months of age. The male schedules focused on economic resources available to men, agricultural practices, household agricultural production, migration, and labor history. A sample of households were chosen for more detailed data on household budgets and expenditures. A sample of land parcels were also chosen. These provided a means to document and monitor agricultural practices, insect infestations, production, and erosion. These approaches were combined with extensive ethnographic techniques. These included the collection of a sample of economically focused life histories in order to document and understand the economic decision making processes of women and men as individuals and within households, the changing economic alternatives available to them, and women's and men's predictions about the future outcomes of the choices they were making.

Because research goals included understanding trends over time, an attempt was made to collect data that spanned a number of years. At the levels of the nation, region, department, and municipality, government publications were most useful. Agricultural censuses from 1952, 1965, and 1974 were consulted (DGECH 1954, 1968, 1976) and then supplemented by data from more recent regional development, natural resource, and agricultural reports, as well as by data from remote sensing devices (aerial photos and satellite imagery). At the local level, additional interviews designed to measure changing use of land resources over the same 30 year period were administered to farmers. These were augmented by monitoring selected parcels of land for a period of one year.

Documented changes in land utilization patterns were tied to social factors influencing the evolution of such patterns and to their ecological outcomes through an integrated master data base. This data base was used first as input to the geographic and agro-economic information system developed by the Comprehensive Resource Inventory Evaluation System Project at Michigan State University (CRIES 1984) and later disaggregated into files managed by dBASE—although other data base managers could have been used. The master data base consisted of physical, ecological, and agricultural information contained in digitized land use, topographic, soil, and land parcel maps; tabular ecological data collected from parcel monitoring; as well as primary data collected through various agricultural, socioeconomic, and nutritional household surveys. The result was a geographically referenced data base keyed to the level of the household but capable of being disaggregated to the levels of the male and female householder or concatenated to the levels of the community, municipality or region. In addition, the identification variables (of individuals, households, communities, etc.) in the master data base were keyed to those in the computer coded ethnographic field notes making it possible to coordinate searches of field notes with analyses of the more "quantified" information included in the master data base.

Altogether the research methodology facilitated overall analysis capable of integrating qualitative and quantitative sources of data. It resulted in a regional land use study that linked social and ecological processes through different levels. The dynamic data base was not only able to answer descriptive questions such as "To what extent has pasture expanded into highland areas?" or "Where and to what degree does burning take place?" but also questions aimed at understanding processes such as "*Why* do farmers continue to burn fields even though they are aware of the negative environmental consequences?" and "*How* does the allocation of household labor influence land use decisions?

Bibliography

ADAI (El Ateneo de la Agroindustria). 1987. *Informe del Seminario: Lineamientos para un Mejor Aprovechamiento de la Ayuda Alimentaria*. Doct no:42/87. Tegucigalpa, Honduras: ADAI.

Adams, Richard N. 1970. *Crucifixion by Power*. Austin, TX: University of Texas Press.

———. 1975. *Energy and Structure: A Theory of Social Power*. Austin, TX: University of Texas Press.

Aguilar, Herling. 1988. *Uso y Efecto de Plaguicidas en Honduras. Informe del Trabajo*. Unidad de Investigación Ambiental, Universidad Nacional Autónoma de Honduras (UNAH). Tegucigalpa, Honduras: UNAH.

Aldenderfer, Mark. 1992. "The State of the Art: GIS and Anthropological Research." *Anthropology Newsletter* 33(5, May): 15.

Aldenderfer, Mark and Herbert Maschner, eds. n.d. *The Anthropology of Human Behavior Through Geographic Information and Analysis*. New York, NY: Oxford University Press.

ANDAH (National Association of Shrimp Farmers of Honduras). 1990. *Shrimp Cultivation: A Positive Support to the Development of Honduras*. Choluteca, Honduras: ANDAH.

Arizpe, L., and J. Aranda. 1986. "Women Workers in the Strawberry Agribusiness in Mexico." In *Women's Work: Development and the Division of Labour by Gender*, Boston, MA: Bergin and Garvey Publishers 173–193.

BANADESA (Banco Nacional de Desarrollo Agrícola). 1987. Préstamos Otorgados por Sistema BANADESA, Boletín Estadístico BANADESA. Tegucigalpa, Honduras. Mimeograph.

Banegas Archaga, Volanda S., Sonia M. Figueroa Cuellar, Yogena R. Paredes Nuñez, and Yoconda Colindres Zuniga. 1991. *La Industria del Camaron en la Zona Sur de Honduras: Su Contribución al Mejoramiento Socioeconomico de la Zona*. M.Sc. Thesis. Universidad Nacional Autónoma de Honduras, Tegucigalpa, Honduras.

BANTRAL (Banco Central de Honduras). 1987. Préstamos por Sistema BANCARIO, Boletín Estadístico BANTRAL - Preliminar. Tegucigalpa, Honduras. Mimeograph.

Barlett, Peggy. 1980. "Adaptive Strategies in Peasant Agricultural Production." *Annual Review of Anthropology* 9: 545–573.

Barry, Tom and Deb Preusch. 1986. *The Central America Fact Book*. New York, NY: Grove Press.

Bartra, Roger. 1974. *Estructura agraria y clases sociales en México*. México, D. F.

Bassett, T. J. 1988. "The Political Ecology of Peasant-herder Conflicts in the Northern Ivory Coast." *Association of American Geographers, Annals* 78(3): 453–472.

Baudez, Claude F. 1966. "Noveau Ceramiques au Honduras: Une Reconsideration de L'evolution Culturalle." *Journal de La Société des Americanistes* LV(2): 299–342.

Behrens, Clifford A. and Thomas L. Sever (eds.). 1991. *Applications of Space-Age Technology in Anthropology*. November 28, 1990 Conference Proceedings. John C. Stennis Space Center, MS: NASA Sciene and Technology Laboratory.

Bennett, John W. 1969. *Northern Plainsmen: Adaptive Strategies and Agrarian Life.* Chicago, IL: Aldine.

———. 1976. *The Ecological Transition: Cultural Anthropology and Human Adaptation.* New York, NY: Pergamon Press.

———. 1980. "Human Ecology as Human Behavior: A Nomative Anthropology of Resource Use and Abuse." In *Human Behavior and Environment,* Vol. 4. edited by I. Altman, A. Rappaport, and J. Wohlwill. New York, NY: Plenum Press.

———. 1985. "The Micro-Macro Nexus: Typology, Process, and Systems." In *Micro and Macro Levels of Analysis in Anthropology: Issues in Theory and Research,* edited by Billie R. DeWalt and P. Pelto. Boulder, CO: Westview Press. 23–54.

———. 1986. "Summary and Critique: Interdisciplinary Research on People-Resources Relations." In *Natural Resources and People: Conceptual Issues in Interdisciplinary Research,* edited by Kenneth A. Dahlberg and John W. Bennett. Boulder, CO: Westview Press. 343–372.

———. 1990. "Ecosystems, Environmentalism, Resource Conservation, and Anthropological Research." In *The Ecosystem Approach in Anthropology: From Concept to Practice,* edited by Emilio F. Moran. Ann Arbor, MI: University of Michigan Press. 435–457.

Benzoni, Girolamo. 1967. *La Historia del Mundo Nuevo.* 1565. Caracas: Biblioteca de la Academia Nacional de la Historia.

Blaikie, Piers. 1985. *The Political Economy of Soil Erosion in Developing Countries.* London: Longman.

———. 1988. "Land Degradation in Nepal." In *Deforestation: Social Dynamics in Watersheds and Mountain Ecosystems,* edited by J. Ives and D. C. Pitt. London: Routledge.

Blaikie, Piers, and Harold Brookfield. 1987. *Land Degradation and Society.* London and New York: Methuen.

Boardman, R. 1986. *Pesticides in World Agriculture.* New York, NY: St. Martin's Press.

Booth, John A., and Thomas W. Walker. 1989. *Understanding Central America.* Boulder, CO: Westview Press.

Boserup, Ester. 1965. *The Conditions of Agricultural Growth.* Chicago, IL: Aldine.

———. 1981. *Population and Technology.* Chicago, IL: University of Chicago Press.

Bottrell, D. 1983. "Social Problems in Pest Management in the Tropics." *Insect Science and Applications* 4(1-2): 179–184.

Boyer, Jefferson C. 1982. *Agrarian Capitalism and Peasant Praxis in Southern Honduras.* Ph.D. dissertation. University of North Carolina, Chapel Hill, NC.

———. 1987. "Capitalism, Campesinos, and Calories in Southern Honduras." In *Directions in the Anthropological Study of Latin America: A Reassessment,* edited by Jack Rollwagen. Albany, NY: SUNY Press.

———. 1991. "Is Sustainable Agriculture and Development Possible in Southern Honduras?" Paper presented at the conference, *Varieties of Sustainability: Reflecting on Ethics, Environment, and Economic Equity,* sponsored by the Agriculture and Human Values Society and the Agroecology Program, University of California-Santa Cruz. Asilomar Conference Center, Pacific Grove, CA, 10–12 May.

Boyer, Jefferson C. and Tammie Church. 1987. "Qué hay en la bodega? The Vanishing Reserves of Peasant Economies in Southern Honduras." Paper presented at the *Conference of the Southeastern Council of Latin American Studies.* Mérida, Yucatán, 1–5 April.

Brand, Charles A. 1972. *The Background of Capitalistic Underdevelopment: Honduras to 1913.* Ph.D. dissertation. University of Pittsburgh, Pittsburgh, PA.

Brandon, Katrina E., and Carter Brandon. 1992. "Linking Environment to Development: Introduction." *World Development* 20(4, April): 477–479.

Brockett, Charles. 1988. *Land, Power, and Poverty: Agrarian Transformation and Political Conflict in Central America*. Boston, MA: Unwin Hyman.

Brown, Lester. 1987. "Analyzing the Demographic Trap." In *State of the World: 1987*, edited by Lester Brown et al., New York, NY: W. W. Norton. 20–37.

Brush, Stephen. 1977. "The Myth of the Idle Peasant." In *Peasant Livelihood*, edited by R. Halperin and J. Dow. New York, NY: St. Martin's Press, 60–78.

———. 1987. "Who Are Traditional Farmers?" In *Household Economies and Their Transformations*, edited by M. D. Machlachlan. Lanham, MD: University Press of America. 143–154.

Bull, D. 1982. *A Growing Problem: Pesticides and the Third World Poor*. Oxford: OXFAM.

Bulmer-Thomas, Victor. 1987. *The Political Economy of Central America Since 1920*. Cambridge, MA: Cambridge University Press.

Buseo, Jose Antonio, Catherine Castañeda, Flora Duarte, and Miriam Chavez. 1987. *Efectos de Plaguicidas en Honduras*. Tegucigalpa, Honduras: Universidad Nacional Autonoma de Honduras.

Cardoso, Fernando H. 1977. "The Consumption of Dependency Theory in the United States." *Latin American Research Review* 12: 7–24.

Carias, Marco Virgilio, and Daniel Slutzky, eds. 1971. *La Guerra Inútil: Análisis Socioeconómico del Conflicto Entre Honduras y El Salvador*. San José, Costa Rica: EDUCA.

Castañeda, C., and Z. Matamoros. 1990. *Environmental Analysis for the Investment and Export Diversification Project*. Tegucigalpa, Honduras: USAID/Honduras.

CCAR (Cotton Cooperative Annual Report). 1978–1982. *Annual Report*. San Lorenzo, Honduras: Cotton Cooperative Annual Report.

CEDOH (Centro de Documentación de Honduras). 1988. 25 Años de Reforma Agraria. Especial No. 34 (Marzo). Tegucigalpa, Honduras: CEDOH.

CELADE/DGECH. 1986. Encuesta demográfica nacional de Honduras. In *FECUNDIDAD*, vol. 4, sec A. San José, Costa Rica: CELADE.

CEPAL (Proyecto de Necesidades Básicas en el Istmo Centroamericano a base de información de los países y CELADE). 1982. *Boletín Demográfico*. 28.

Chapman, Anne M. 1960. "Los Nicarao y Los Chorotega Segun las Fuentes Históricas." In *Serie Historia y Geografía*, vol. 4. San José, Costa Rica: Universidad de Costa Rica.

Chapman, M. D. 1989. "The Political Ecology of Fisheries Depletion in Amazonia." *Environmental Conservation* 16(4): 331–337.

Clark, C. 1974. "The Economics of Overexploitation." *Science* 181: 630–634.

Cloud, David S. 1990. "Who Should Get U.S. Food Aid? Congress Wants to Decide." *Congressional Quarterly Weekly Report* 48(35, September 1): 2767–2770.

Collins, Jane. 1986. "Smallholder Settlement of Tropical South America: The Social Causes of Ecological Destruction." *Human Organization* 45(1): 1–10.

———. 1987. "Labor Scarcity and Ecological Change." In *Lands at Risk in the Third World*, edited by Peter D. Little and Michael M. Horowitz. Boulder, CO: Westview Press. 19–37.

———. 1988. *Unseasonal Migrations: The Effects of Rural Labor Scarcity in Peru*. Princeton, NJ: Princeton University Press.

CONAMA (Comisión Nacional del Medio Ambiente y Desarrollo). 1991. *Honduras Environmental Agenda*. Tegucigalpa, Honduras: Graficentro Editores.

Conant, Francis. 1990. "1990 and Beyond: Satellite Remote Sensing and Ecological Anthropology." In *The Ecosystem Approach in Anthropology: From Concept to Practice*, edited by Emilio F. Moran. Ann Arbor, MI: University of Michigan Press. 357–388.

———. 1992. "Thoughts on GIS and Their Implications for Anthropology." *Anthropology Newsletter* 33(5, May): 15.

CONSUPLANE (Secretaria Tecnica del Consejo Superior de Planificacion Economica). 1982. *Las Regiones: Planificacion.* Tegucigalpa, Honduras: CONSUPLANE.

Corbett, Jack. 1991. "Food Security and Regional Development." In, *Harvest of Want: Food Security in Central America and Mexico,* edited by Anne Ferguson and Scott Whiteford. Boulder, CO: Westview Press, 243–254.

CRIES (Comprehensive Resource Inventory and Evaluation System Project). 1984. *Resource Assessment of the Choluteca Department.* East Lansing, MI: Michigan State University and the United States Department of Agriculture.

CSM (*Christian Science Monitor*). 1990. "Encourage Honduras in its Bold Economic Reforms." *Christian Science Monitor* 30 April. p. 107.

CSPE/OEA (Secretaria Técnica del Consejo Superior de Planificación Económica y Secretaria General de la Organización de Estados Americanos). 1982. *Proyecto de Desarrollo Local del Sur de Honduras.* Tegucigalpa, Honduras: CSPE.

Danaher, Kevin, Phillip Berryman, and Medea Benjamin. 1987. "Help or Hindrance: United States Economic Aid in Central America." *Food First Development Report No. 1.* San Francisco, CA: the Institute for Food and Development Policy.

Danes, Sharon M., Mary Winter, and Michael B. Whiteford. 1987. "Level of Living and Participation in the Informal Market Sector Among Rural Honduran Women." *Journal of Marriage and the Family* 49(August): 631–639.

de Janvry, Alain. 1981. *The Agrarian Question and Reformism in Latin America.* Baltimore, MD: The John Hopkins University Press.

de Janvry, Alain, and Ann Vandeman. 1987. "Patterns of Proletarianization in Agriculture: an International Comparison." In *Household Economies and Their Transformations,* edited by M. D. Maclachlan, Lanham, MD: University Press of America. 28–73.

Deere, Carmen Maria, editor 1987. *Rural Women and State Policy: Feminist Perspectives on Latin American Agricultural Development.* Boulder, CO: Westview Press.

———. 1990. *Household and Class Relations: Peasants and Landlords in Northern Peru.* Berkeley, CA: University of California Press.

Deere, Carmen Diana, and Alain de Janvry. 1979. "A Conceptual Framework for the Empirical Analysis of Peasants." *American Journal of Agricultural Economics* 61(4, November): 601–611.

Deere, Carmen, and Robert Wasserstrom. 1981. "Ingreso Familiar y Trabajo no Agrícola entre los Pequeños Productores de América Latina y El Caribe." In *Agricultura de Ladera en América Tropical: Informe Técnico No. 11,* edited by A. R. Novoa and J. L. Posner, Turrialba, Costa Rica: CATIE. 151–167.

Denevan, William M., ed. 1976. *The Native Population of the Americas in 1492.* Madison, WI: University of Wisconsin Press.

———. 1992. *The Native Population of the Americas in 1992,* 2nd edition. Madison, WI: University of Wisconsin Press.

DeWalt, Billie R. 1979. *Modernization in a Mexican Ejido.* New York, NY: Cambridge University Press.

———. 1985. "Microcosmic and Macrocosmic Processes of Agrarian Change in Southern Honduras: The Cattle Are Eating the Forest." In *Micro and Macro Levels of Analysis in Anthropology: Issues in Theory and Research,* edited by Billie R. DeWalt and P. J. Pelto, Boulder, CO: Westview Press. 165–186.

DeWalt Billie R. and Kathleen M. DeWalt. 1982. *Socioeconomic Constraints in the Production, Distribution and Consumption of Sorghum in Southern Honduras.* INTSORMIL, Farming Systems Research in Southern Honduras. Report No. 1. Lexington, KY: University of Kentucky Department of Anthropology.

DeWalt, Kathleen M. and Billie R. DeWalt. 1987. "Nutrition and Agricultural Change in Southern Honduras." *Food and Nutrition Bulletin* 9(3): 36–45.

DeWalt Billie R., and Susan C. Stonich. 1985. "Farming Systems Research in Southern Honduras." In *Fighting Hunger with Research: A Five-Year Technical Research Report of the Grain Sorghum/Pearl Millet Collaborative Research Support Program*, edited by Judy F. Winn. Lincoln, NE: University of Nebraska, 184–192.

DGECH (Dirección General de Estadística y Censos). 1954. *Censo Nacional Agropecuario 1952*. Tegucigalpa, Honduras: DGECH.

_____. 1968. *Censo Nacional Agropecuario 1965*. Tegucigalpa, Honduras: DGECH.

_____. 1976. *Censo Nacional Agropecuario 1974*. Tegucigalpa, Honduras: DGECH.

_____. 1981. *Censos de Población y Vivienda Levantados en Honduras de 1791 a 1974*. Tegucigalpa, Honduras: DGECH.

Dove, Michael R. 1983. "Theories of Swidden Agriculture and the Political Economy of Ignorance." *Agroforestry Systems* 1: 85–99.

Durham, William H. 1979. *Scarcity and Survival in Central America: The Ecological Origins of the Soccer War*. Stanford, CA: Stanford University Press.

Eckholm, E. P. 1976. *Losing Ground: Environmental Stress and World Food Prospects*. New York, NY: Norton.

_____. 1982. *Down to Earth*. London: Pluto Press.

ECLA (Economic Comission for Latin America). 1955. *Economic Survey of Latin America*, New York, NY: United Nations. 117–119.

ECLAC (Economic Commission for Latin America and the Caribbean). 1986. Central America: "Bases for a Reactivation and Development Policy." *CEPAL Review* 28(April): 11–48.

_____. 1987. *Statistical Yearbook for Latin America and the Caribbean*. New York, NY: United Nations.

_____. 1989. *Preliminary Overview of the Economy of Latin America and the Caribbean*. Santiago, Chile: UN: ECLAC.

_____. 1990. *The Water Resources of Latin America and the Caribbean—Planning, Hazards, and Pollution*. Santiago, Chile: UN/ECLAC.

Ehrlich, Paul R. 1968. *The Population Bomb*. New York, NY: Ballantine Books.

Ehrlich, Paul R., and Ann H. Ehrlich. 1970. *Population, Resources, Environment: Issues in Human Ecology*. San Francisco, CA: W. H. Freeman.

_____. 1990. *The Population Explosion*. New York, NY: Simon and Schuster.

Ellen, R. 1982. *Environment, Subsistence, and System*. Cambridge, MA: Cambridge University Press.

FAO (Food and Agricultural Organization of the United Nations). 1984. *Food Balance Sheets: 1979–81 Average*, 103–4. Rome: FAO.

FAO-PY (Food and Agricultural Organization of the United Nations). Various years. *Production Yearbook*. Rome: United Nations.

FAO-TY (Food and Agricultural Organization of the United Nations). Various years. *Trade Yearbook*. Rome: United Nations.

Faber, Daniel. 1992a. "The Ecological Crisis of Latin America." *Latin American Perspectives* 19(1, Winter): 3–16.

_____. 1992b. "Imperialism, Revolution, and the Ecological Crisis of Central America." *Latin American Perspectives* 19(1, Winter): 17–44.

Favre, Henri. 1977. "The Dynamics of Indian Peasant Society and Migration to Coastal Plantations in Central Peru." In *Land and Labour in Latin America*, edited by Kenneth Duncan and Ian Rutledge, Cambridge: Cambridge University Press. 83–102.

Feeny, David, Fikret Berkes, Bonnie McCay, and James Acheson. 1990. "The Tragedy of the Commons: Twenty-Two Years Later." *Human Ecology* 18(1): 1–19.

Finney, Kenneth. 1973. *Precious Metal Mining and the Modernization of Honduras: In Quest of El Dorado*. Ph.D. dissertation. Tulane University. New Orleans, LS.

Fordham, Miriam, Billie R. DeWalt, and Kathleen M. DeWalt. 1985. *The Economic Role of Women in a Honduran Peasant Community.* Farming Systems Research in Southern Honduras, Report No. 2. Lexington, KY: Department of Sociology and Department of Anthropology, University of Kentucky.

Forster, Nancy R. 1992. "Protecting Fragile Lands: New Reasons to Tackle Old Problems." *World Development* 20(4): 571–585.

Garcia, Magdalena, Roger D. Norton, Mario Ponce, and Roberta van Haeften. 1988. *Agricultural Development Policies in Honduras: A Consumption Perspective.* Tegucigalpa, Honduras: USDA/USAID.

Garst, Rachel and Tom Barry. 1990. *Feeding the Crisis: U.S. Food Aid and Farm Policy in Central America.* Lincoln and London: University of Nebraska Press.

Goldrich, Daniel and David V. Carruthers. 1992. "Sustainable Development in Mexico? The International Politics of Crisis or Opportunity." *Latin American Perspectives* 19(1): 97–122.

Gonzalez, J. 1987. *Situación de la Carcinocultura en la Costa Sur de Honduras.* Tegucigalpa, Honduras: RENARE.

Goodchild, Michael. 1992. "GIS in Anthropology: Where Do We Stand?" *Anthropology Newsletter* 33(5, May): 14–15.

Grosh, Margaret. 1990. *Social Spending in Latin America: The Story of the 1980s.* World Bank Discussion Papers, no. 106. Washington, D.C.: World Bank.

Grossman, Larry. 1984. "Cattle, Rural Economic Differentiation, and Articulation in the Highlands of Papua, New Guinea." *American Ethnologist* 10: 59–76.

Hansen, M. 1990. *The First Three Years: Implementation of the World Bank Pesticide Guidelines, 1985-1988.* Washington, D.C.: Consumers Union.

Hardin, Garrett J. 1968. "The Tragedy of the Commons." *Science* 162: 1243–1248.

——. 1977. *The Limits to Altruism: An Ecologist's View of Survival.* Bloominton, IN: University of Indiana Press.

Hardin, Garrett, and John Baden, eds. 1977. *Managing the Commons.* San Francisco, CA: W. H. Freeman.

Hargreaves, George. 1980. *Monthly Precipitation Probabilities for Moisture Availability for Honduras.* Logan, UT: Utah State University.

Hawkins, Richard. 1984. *Intercropping Maize with Sorghum in Central America: A Cropping System Case Study.* Turrialba, Costa Rica: Centro Agrónomico Tropical de Investigación y Enseñanza (CATIE).

Hayami, Yujiro and Vernon Ruttan. 1985. *Agricultural Development: An International Perspective.* Baltimore, MD: Johns Hopkins University Press.

Hecht, Susanna. 1981. "Cattle Ranching in Amazonia: Political and Ecological Considerations." In *Frontier Expansion in Amazonia,* edited by M. Schmink and C. Wood, Gainesville, FL: University of Florida Press. 366–386.

Heffernan, K. 1988. "Problems and Prospects of Export Diversification: Honduras." In *Struggle Against Dependence: Nontraditional Export Growth in Central America and the Caribbean,* edited by E. Paus, Boulder, CO: Westview Press. 123–144.

Heynig, Klaus. 1982. "The Principal Schools of Thought on the Peasant Economy." *CEPAL Review* 16(April): 113–140.

Holdridge, L. R. 1962. *Mapa Ecológico de Honduras* San José, Costa Rica: Tropical Science Center.

Holloway, Marguerite. 1992. "Population Pressure: The Road from Rio is Paved with Factions." *Scientific American* 267(3, September): 32, 36, 38.

Holmes, Douglas. 1983. "A Peasant-Worker Model in a Northern Italian Context." *American Ethnologist* 10: 734–748.

Honduras This Week. 1992. "Planned Construction of Airport for City of Choluteca." *Honduras This Week* 5(26, July 11): 1, 19.

Howard Ballard, Patricia. 1987. *From Banana Republic to Cattle Republic: Agrarian Roots of the Crisis in Honduras*. Unpublished Ph.D. dissertation. Ann Arbor, MI: University Microfilms.

Howard-Borjas, Patricia. 1990. *Empleo y pobreza rurales en Honduras, con enfoque especial en la mujer*. Mimeograph. Honduras: SECPLAN/PNUD.

ICAITI (Central American Institute of Investigation and Industrial Technology). 1977. *An Environmental and Economic Study of the Consequences of Pesticide Use in Central American Cotton Production*. Washington, D.C.: USAID.

IHMA (Instituto Hondureño de Mercadeo Agrícola). 1987. *Análisis y Propuestas para el Establecimiento de los Precios de Garantía para Granos Básicos*. Tegucigalpa, Honduras: IHMA.

INCAP (Instituto de Nutrición de Centro América y Panamá). 1969. *Evaluación Nutricional de la Poblacion de Centro América y Panamá: Honduras*. Guatemala City, Guatemala: Instituto de Nutrición de Centro América y Panamá.

Ives, Jane H., ed. 1985. *The Export of Hazard: Transnational Corporation and Environmental Control Issues*. Boston, MA: Routledge & Kegan Paul. 94–114.

Jacobson, Jodi L. 1992. "Improving Women's Reproductive Health." In *State of the World: 1992*, edited by Lester Brown et al. New York, NY: W. W. Norton & Co.

Janzen, Daniel. 1973. "Tropical Agro-Ecosystems." *Science* 182: 1213–1218.

Jarvis, Lovel S. 1986. *Livestock Development in Latin America*. Washington, DC: World Bank.

Johannessen, Carl L. 1963. *Savannas of Interior Honduras*. Berkeley, CA: University of California Press.

Johnson, D. H. 1989. "Political Ecology in the Upper Nile: Twentieth Century Expansion of the Pastoral Common Economy." *Journal of African History* 30(3): 463–486.

Jones, Jeffrey R. 1982. *Producción y Consumo de Leña en Fincas Pequenas de Honduras*. Turrialba, Cost Rica: Centro Agrónomico Tropical de Investigación y Enseñanza (CATIE).

Karliner, Joshua. 1989. "The Ecological Destabilization of Central America." *World Policy Journal* 7(Fall): 787–810.

Keppner, Charles D. Jr. 1936. *Social Aspects of the Banana Industry*. New York, NY: Columbia University Press.

Keppner, Charles D. Jr., and Jay Soothill. 1935. *The Banana Empire: A Case Study of Economic Imperialism*. New York, NY: Vanguard Press.

Krueger, C. 1989. "Development and Politics in Rural Guatemala." *Development Anthropology Network: Bulletin of the Institute for Development Anthropology* 7(1): 1–6.

LaBarge, Richard A. 1959. *A Study of United Fruit Company Operations in Isthmian America 1946–1956*. Ph.D. dissertation. North Carolina: Duke University. Durham, NC.

LACR (Latin American Commodities Report). 1986. "Meat / Honduras." *Latin American Commodities Report*. CR-86-08. 25 April.

_____. 1989a. "Shrimp / Central America." *Latin American Commodities Report*. CR-89-05. 15 May.

_____. 1989b. "Shrimp / Honduras." *Latin American Commodities Report*. CR-89-07. 15 July.

_____. 1989c. "Fruit / Honduras." *Latin American Commodities Report*. CR-89-07. 15 July.

LAEB (Latin American Economy and Business). 1991. "Rising rates and tight credit hit small farmers." *Latin American Economy and Business*. LAEB-91-10. October.

Lainez, Vilma, and Victor Meza. 1973. "El Enclave Bananero en la Historia de Honduras." *Estudios Sociales Centroamericanos* II(5): 141–148.

Lappé, Francis Moore, and Rachel Schurman. 1988. "The Missing Piece in the Population Puzzle." *Food First Development Report No. 4*. San Francisco, CA: The Institute for Food and Development Policy.

LARR (Latin American Regional Report). 1991a. "Honduras/IMF." *Latin American Regional Report*. RM-91-06. 18 July.

———. 1991b. "Inflation and exchange rates in the region." *Latin American Regional Report*. RM-91-07. 22 August.

———. 1992a. "Inflation and exchange rates in the region." *Latin American Regional Report*. RM-92-02. 26 March.

———. 1992b. "Update." *Latin American Regional Report*. RM-92-04. 7 May.

———. 1992c. "Update." *Latin American Regional Report*. RM-92-05. 11 June.

———. 1992d. "Finance." *Latin American Regional Report*. RM-92-07. 20 August.

Larson, David. 1982. "The Problems and Effects of Price Controls on Honduran Agriculture." Mimeograph. Minneapolis, MN: Experience Incorporated.

Las Casas, Bartolomé de. 1957-58. *Biblioteca de Autores Españoles*. Vol. 95, 96, 105, 106, 110, *Obras Escogidas de Fray Bartolomé de las Casas*. edited by Juan Pérez de Tudela.

LaFeber, Walter. 1983. *Inevitable Revolutions: The United States in Central America*. New York, NY: W. W. Norton.

Leff, Enrique. 1986. *Ecología y capital: Hacía una perspectiva ambiental del desarrollo*. Mexico City: Universidad nacional Autónoma de México.

Lehmann, David. 1982. "After Chayanov and Lenin: New Paths of Agrarian Capitalism." *Journal of Development Economics* 11: 133–161.

Lele, Uma and Steven W. Stone. 1989. "Population Pressure, the Environment and Agricultural Intensification: Variations on the Boserup Hypothesis." *MADIA Discussion Paper* 4. Washington, D.C.: World Bank.

Leonard, Jeffrey H. 1987. *Natural Resources and Economic Development in Central America*. New Brunswick, NJ: Transaction Books.

Little, Peter, and Michael Horowitz. 1987. "Social Science Perspectives on Land, Ecology, and Development." In *Lands at Risk in the Third World: Local Level Perspectives*, edited by P. Little and M. Horowitz, Boulder, CO: Westview. 1–16.

López, José Gabriel. 1990. *Agrarian Transformation and the Political, Ideological, and Cultural Responses from the Base: A Case Study from Western Mexico*. Ph.D. dissertation. University of Texas, Austin, TX.

Lowe, Justin. 1992. "The Road from Rio: Where Do We Go from From Here?" *Earth Island Journal* 7(3): 41–42.

Maclachlan, Morgan D. 1987. "From Intensification to Proletarianization." In *Household Economies and their Transformations: Monographs in Economic Anthropology, No. 3*, edited by M. D. Maclachlan, Lanham, MD: University Press of America. 1–27.

MacLeod, Murdo J. 1973. *Spanish Central America: A Socioeconomic History, 1520–1720*. Berkeley, CA: University of California Press.

Marinas Otero, Luis. 1963. *Honduras*. Madrid: Ediciones Cultura Hispana.

McCay, Bonnie J., and Acheson James. 1987. *The Culture and Ecology of Communal Resources*. Tucson, AZ: University of Arizona Press.

McBride, Tim. 1992. "Research Agreement Helps Balance Food, Resource Needs in Honduras." *Front Lines* May: 9.

Meckenstock, Dan, David Coddington, Juan Rosas, Harold van Es, Manjeet Chinman, and Manuel Murillo. 1991. "Honduras Concept Paper: Towards a Sustainable Agriculture in Southern Honduras." Paper presented at the International Sorghum/Millet Collaborative Research Support Conference, June 8–12, Corpus Christi, TX.

Mejia, R. 1991. *El Sector de Camaron en la Programa de Ajuste Estructural*. M.Sc. Thesis. Universidad Nacional Autónoma de Honduras, Tegucigalpa, Honduras.

Messer, Judy Anne. 1987. *The Political Ecology of Agriculture: The Case Study of Australia*. Ph.D. dissertation. University of New South Wales, Australia.

Mintz, Sidney. 1977. "The So-called World System: Local Initiative and Local Response." *Dialectical Anthropology* 2(4): 253–270.

Molina, Guillermo. 1975. "Población, estructura productiva y migraciones internas en Honduras: 1950–1960." *Revista de Estudios Sociales Centroamericanos*, 4(12): 9–39.

Moran, Emilio F. 1986. "Anthropological Approaches to the Study of Human Impacts." In *Natural Resources and People: Conceptual Issues in interdisciplinary Research*, edited by Kenneth A. Dahlberg and John W. Bennett. Boulder, CO: Westview Press. 107–128.

———. 1990. "Levels of Analysis and Analytical Level Shifting: Examples from Amazonian Ecosystem Research." In *The Ecosystem Approach in Anthropology: From Concept to Practice*. edited by Emilio F. Moran. Ann Arbor, MI: University of Michigan Press. 279–308.

———. 1992. "Anthropology and Remote Sensing." *Culture and Agriculture* 43(Spring): 16–17.

MSP (Ministerio de Salud Pública). 1989. *Encuesta nacional de nutrición, 1987*. Tegucigalpa, Honduras: MSP.

Murdoch, William. 1980. *The Poverty of Nations*. Baltimore and London: The Johns Hopkins University Press.

Murray, Douglas. 1991. "Export Agriculture, Ecological Disruption, and Social Inequality: Some Effects of Pesticides in Southern Honduras." *Agriculture and Human Values* 8(4, Fall): 19–29.

Murray, Douglas, and Polly Hoppin. 1990. "Pesticides and Nontraditional Agriculture: A Coming Crisis for US Development Policy in Latin America." In *Texas Papers on Latin America*. Paper no. 90-04. Austin, TX: Institute of Latin American Studies, University of Texas, Austin.

———. 1992. "Recurring Contradictions in Agrarian Development: Pesticides and Caribbean Basin Nontraditional Agriculture." *World Development* 20(4, April): 597–608.

Myers, N. 1981. "The Hamburger Connection: How Central America's Forests Become North America's Hamburgers." *Ambio* 10(1): 3–8.

Nash, June. 1986. "A Decade of Research on Latin American Women." In *Women and Change in Latin America*, South Hadley, MA: Bergin and Garvey Publishers, 3–21.

National Research Council (NRC). 1982. *The Ecological Aspects of Development in the Humid Tropics*. Washington, D.C.: National Academy Press.

———. 1986. *Population Growth and Economic Development: Policy Questions*. Washington, D.C.: National Academy Press.

———. 1992. *Global Environmental Change: Understanding the Human Dimension*. Washington, D.C.: National Academy Press.

Nations, James, and D. Komer. 1983. "Rainforests and the Hamburger Society." *Environment* 25(3): 12–20.

Nations, James, and Robert Nigh. 1980. "The Evolutionary Potential of Lacandon Maya Sustained Yield Tropical Forest Agriculture." *Journal of Anthropological Research* 36: 1–30.

Netting, Robert. 1968. *Hill Farmers of Nigeria*. Seattle, WA: University of Washington Press.

———. 1981. *Balancing on an Alp*. New York, NY: Cambridge University Press.

Norton, Roger and Carlos Benito. 1987. *An Evaluation of the PL 480 Title I Program in Honduras*. Winrock International: USAID/Honduras.

O'Brien Fonck, Carlos. 1972. *Modernity and Public Policies in the Context of the Peasant Sector: Honduras as a Case Study*. Cornell University. Ithaca, NY: Latin American Studies Program Dissertation Series.

OAS (Organization of American States). 1962. *Informe Oficial de la Misión 105 de Asistencia Térmico Directa a Honduras Sobre Reforma Agraria y Desarrollo Agrícola. Vol. I.* Washington, D.C.: Organization of American States.

O'Connor, James. 1988. "Capitalism, Nature, Socialism: A Theoretical Introduction." *Capitalism, Nature, Socialism: A Journal of Socialist Ecology* 1(Fall): 11–38.

———. 1989. "Uneven and Combined Development and Ecological Crisis: A Theoretical Introduction." *Race and Class* 30(3): 1–13.

Ortner, Sherry B. 1984. "Theory in Anthropology Since the Sixties." *Comparative Studies in Society and History* 26(1): 126–166.

Oviedo y Valdes, Gonzalo Fernández de. 1959. *Biblioteca de Autores Españoles.* Vol. 117-121, *Historia General y Natural de las Indias.* edited by Juan Pérez de Tudela. Madrid: Ediciones Atlas.

PAHO (Pan American Health Organization). 1990. *Health Conditions in the Americas* Scientific Publication No. 524. Volumes I and II. Washington, D.C.: PAHO.

Painter, Michael. 1987. "Unequal Exchange: the Dynamics of Settler Impoverishment and Environmental Destruction in Lowland Bolivia." In *Lands at Risk in the Third World: Local Level Perspectives,* edited by P. Little and M. Horowitz, Boulder, CO: Westview Press. 169–192.

Parsons, Kenneth H. 1975. *Agrarian Reform in Southern Honduras.* Research Paper No. 67. Madison, WI: Land Tenure Center, University of Wisconsin.

Parsons, James. 1976. "Forest to Pasture: Development or Destruction?" *Revista de Biología Tropical* 24(Supplement 1): 121–138.

Paus, E., ed. 1988. *Struggle Against Dependence: Nontraditional Export Growth in Central America and the Caribbean.* Boulder, CO: Westview Press.

Paz, Ernesto. 1986. "The Foreign Policy and National Security of Honduras." In, *Honduras Confronts Its Future: Contending Perspectives on Critical Issues,* edited by Mark B. Rosenberg and Philip L. Shepherd. Boulder, CO: Lynne Reinner Publishers. 181–210.

Pelto, Perti J. and Billie R. DeWalt, eds. 1985. *Micro and Macro Levels of Analysis in Anthropology: Issues in Theory and Research.* Boulder, CO: Westview Press.

Perez Brignoli, Hector. 1973a. "Economía y Sociedad en Honduras Durante el Siglo XIX." *Estudios Sociales Centroamericanos* II(6): 51–82.

———. 1973b. "La Reforma Liberal in Honduras." *Cuadernos de Ciencias Sociales* 2, 1-135. Tegucigalpa, Honduras: Editorial Nuevo Contiente.

Pingali, Prabhu, Yves Bigot and Hans Binswanger. 1987. *Agricultural Mechanization and the Evolution of Farming Systems in Sub-Saharan Africa.* Washington, D.C.: Johns Hopkins University Press.

Ponce, Mario. 1986. "Honduras: Agricultural Policy and Perspectives." In *Honduras Confronts Its Future: Contending Perspectives on Critical Issues,* edited by Mark B. Rosenberg and Philip Shepherd. Boulder, CO: Lynne Rienner Publishers. 129–152.

Poole, Peter. 1989. "Developing a Partnership of Indigenous Peoples, Conservationists, and Land Use Planners in Latin America." *Latin America and the Carribean Technical Department Working Paper # WPS* 245. New York, NY: The World Bank.

Posas, Mario. 1980. "Honduras at the Crossroads." *Latin American Perspectives.* 7(25–26): 45–56.

Posner, J. L., and M. F. MacPherson. 1982. "Agriculture on the Steep Slopes of Tropical America: The Current Situation and Prospects." *World Development* 10(May): 341–354.

Rappaport, Roy A. 1967. *Pigs for the Ancestors.* New Haven, CN: Yale University Press.

———. 1990. "Ecosystems, Populations and People." In *The Ecosystem Approach in Anthropology: From Concept to Practice,* edited by Emilio F. Moran. Ann Arbor, MI: University of Michigan Press. 41–72.

Redclift, Michael. 1984. *Development and Environmental Crisis.* New York, NY: Methuen.

———. 1987. *Sustainable Development: Exploring the Contradictions.* London and New York: Methuen.

Repetto, Robert. 1992. "Accounting for Environmental Assets." *Scientific American* (June): 94–100.

Roseberry, William. 1983. *Coffee and Capitalism in the Venezuelan Andes*. Austin, TX: University of Texas Press.

———. 1989. *Anthropologies and Histories: Essays in Culture, History, and Political Economy*. New Brunswick and London: Rutgers University Press.

Rosset, Peter M. 1991. "Sustainability, Economies of Scale, and Social Instability: Achilles Heel of Non-Traditional Export Agriculture?" *Agriculture and Human Values* 8(4, Fall): 30–37.

Rubio Melhado, Adolfo. 1953. *Geografía General de la República de Honduras*. Tegucigalpa, Honduras: Ministerio de Educación Pública Calderón.

Ruhl, J. Mark. 1984. "Agrarian Structure and Political Stability in Honduras." *Journal of Interamerican Studies and World Affairs* 26(February): 33–68.

———. 1987. "The Honduran Agrarian Reform Under Suazo Cordova, 1982-85; an Assessment." *Inter-American Economic Affairs* 39(2): 63–80.

SAEH/INCAP (Secretaría de Educación Pública. Dirección General de Educación Primaria. Servicio de Alimentación Escolar de Honduras/Instituto de Nutrición de Centro América y Panamá). 1987. *Primer Censo Nacional de Talla en Escolares de Primer Grado de Educacion Primaria de la República de Honduras, 1986*. Tegucigalpa, Honduras: SAEH/INCAP.

Safa, Helen. 1986. "Runaway Shops and Female Employment: The Search for Cheap Labor." In *Women's Work: Development and the Division of Labour by Gender*. Boston, MA: Bergin and Garvey Publishers, Inc. 58–74.

———. 1987. "Urbanization, the Informal Economy and State Policy in Latin America." In *The Capitalist City*, edited by Michael P. Smith and Joseph Feagin. New York, NY: Basil Blackwell.

SALA (Statistical Abstract of Latin America). 1984. *Statistical Abstract of Latin America*. Vol. 23. edited by J. Wilkie, D. E. Lorey and E. Ochoa. Los Angeles, CA: UCLA Latin American Center Publications, University of California.

———. 1988. *Statistical Abstract of Latin America*. Vol. 26. edited by J. Wilkie, D. E. Lorey and E. Ochoa. Los Angeles, CA: UCLA Latin American Center Publications, University of California.

Saldanha, I. M. 1990. "The Political Ecology of Traditional Farming Practices in Thana Mararashtra, India." *Journal of Peasant Studies* 17(3): 433–443.

Sánchez, Pedro, and Steven Buol. 1975. "Soils of the Tropics and the World Food Crisis." *Science* 188: 598–603.

SAPLAN (Sistema de Análisis y Planificación de la Alimentación y Nutrición). 1981. "Analisis de la Situacion Nutricional Durante el Periodo 1972-1979." Tegucigalpa, Honduras: Sistema de Análisis y Planificación de la Alimentación y Nutrición. Mimeograph.

Schmink, Marianne, and Charles Wood. 1987. "The Political Ecology of Amazonia." In *Lands at Risk in the Third World: Local Level Perspectives*, edited by Peter D. Little and Michael M. Horowitz. Boulder, CO: Westview Press, 38–57.

Schreiner, Dean F., and Dan D. Badger. 1983. *Proposed Long Range Strategy for the Development of Southern Honduras*. International Development Series No. 83-3. Stillwater, OK: Department of Agricultural Economics, Oklahoma State University.

SECPLAN (Secretaría de Planificación, Coordinación y Presupuesto). 1988. *Censo Nacional de Población y Vivienda 1988* (Resultados Preliminares). Tegucigalpa, Honduras: SECPLAN.

———. 1989. *Censo Nacional de Población y Vivienda 1988*. Tegucigalpa, Honduras: SECPLAN.

———. 1990. *Encuesta Agrícola Nacional 1989*, Volumen I. Tegucigalpa, Honduras: SEC-PLAN, May.

SECPLAN/USAID (Secretaria de Planificación, Coordinación, y Presupuesto/United States Agency for International Development). 1989. *Perfil Ambiental de Honduras 1989*. Tegucigalpa, Honduras: USAID.

Seligson, Mitchell A. 1980. *Peasants of Costa Rica and the Development of Agrarian Capitalism*. Madison, WI: University of Wisconsin Press.

Sellers, Stephen G. 1984. "Diet Patterns and Nutritional Intake in a Costa Rican Community." *Ecology of Food and Nutrition* 14: 205–218.

Shane, D. R. 1986. *Hoofprints in the Forest: Cattle Ranching and the Destruction of Latin America's Tropical Forest*. Philadelphia, PA: Institute for the Study of Human Issues.

Shepherd, Philip L. 1985. "The Tragic Course and Consequences of U.S. Policy in Honduras." *World Policy Journal* 2(1): 109–154.

Sheridan, T. E. 1988. *Where the Dove Calls: The Political Ecology of a Peasant Corporate Community in Northwestern Mexico*. Tucson, AZ: University of Arizona Press.

Simon, Julian. 1980. "Resources, Population, Environment: An Oversupply of False Bad News." *Science* 208: 1431–37.

———. 1981. *The Ultimate Resource*. Princeton, NJ: Princeton University Press.

Simon, Julian and Herman Kahn, eds. 1984. *The Resourceful Earth*. Oxford: Basil Blackwell.

Smith, Carol. 1980. "Beyond Dependency Theory: National and Regional Patterns of Underdevelopment in Guatemala." *American Ethnologist* 5: 574–617.

Smith, Gavin. 1989. *Livelihood and Resistance*. Berkeley, CA: University of California Press.

Smith, Joan, Immanuel Wallerstein, and Hans-Dieter Evers, eds. 1984. *Households and the World Economy*. Beverly Hills, CA: Sage.

Smith, N., and P. O'Keefe. 1980. "Geography, Marx, and the Concept of Nature." *Antipode* 12: 2.

Smith, Sheldon and Edward Reeves, eds. 1989. *Human Systems Ecology: Studies in the Integration of Political Economy, Adaptation, and Socionatural Regions*. Boulder, CO: Westview Press.

Speth, James Gustave. 1992. "A Post-Rio Compact." *Foreign Policy* 88 (Fall): 145–161.

Spooner, Brian. 1987. "Insiders and Outsiders in Baluchistan: Western and Indigenous Perspectives on Ecology and Development." In *Lands at Risk in the Third World: Local Level Perspectives*, edited by P. Little and M. Horowitz, Boulder, CO: Westview Press. 58–68.

SRN (Secretaría de Recursos Naturales). 1982. *Los Granos Básicos en su Aspecto Económico*. Programa Nacional de Granos Básicos, Dirección Agrícola, Regional del Sur. Tegucigalpa, Honduras: SRN.

Stares, Rodney. 1972. *La Economía Campesina en la zona Sur de Honduras: 1950–1970*. Choluteca, Honduras: Prepared for the Bishop of Choluteca.

Stavenhagen, Rodolfo. 1977. "Basic Needs, Peasants, and the Strategy for Rural Development." In *Another Development: Approaches and Strategies*, edited by M. Nerfin. Uppsala.

———. 1978. "Capitalism and the Peasantry in Mexico." *Latin American Perspectives* 5: 27–37.

Stokes, William S. 1950. *Honduras: An Area Study in Government*. Madison, WI: University of Wisconsin Press.

Stone, Doris Z. 1957. *Papers of the Peabody Museum of Archaelogy and Ethnology: The Archaelolgy of Central and Southern Honduras*. Vol. 49, 3. Cambridge, MA: Harvard University.

Stonich, Susan C. 1986. *Development and Destruction: Interrelated Ecological, Socioeconomic, and Nutritional Change in Southern Honduras*. Ph.D. dissertation. University of Kentucky, Lexington, KY.

———. 1989. "The Dynamics of Social Processes and Environmental Destruction: A Central American Case Study." *Population and Development Review* 15(2): 269–296.

———. 1991a. "Rural Families and Income From Migration: Honduran Households in the World Economy." *Journal of Latin American Studies* 23(1): 131–161.

———. 1991b. "The Political Economy of Food Security in Honduras." In *Harvest of Want: Food Security in Central America and Mexico*, edited by Anne Ferguson and Scott Whiteford. Boulder, CO: Westview Press, 45–74.

———. 1991c. "The Promotion of Nontraditional Exports in Honduras: Issues of Equity, Environment, and Natural Resource Management." *Development and Change* 22(4): 725–755.

———. 1992. "Struggling with Honduran Poverty: The Environmental Consequences of Natural Resource Based Development and Rural Transformation." *World Development* 20(3, March): 385–399.

———. n.d. "Integrating Socioeconomic and Geographic Information Systems: A Methodology for Rural Development and Agricultural Policy Design." In *The Anthropology of Human Behavior Through Geographic Information and Analysis*, edited by Mark Aldenderfer and Herbert Maschner. New York, NY: Oxford University Press.

Stonich, Susan C., and Billie R. DeWalt. 1989. "The Political Economy of Agricultural Growth and Rural Transformation in Honduras and Mexico." In *Human Systems Ecology: Studies in the Integration of Political Economy, Adaptation, and Socionatural Regions*, edited by Sheldon Smith and Edward Reeves. Boulder, CO: Westview Press, 202–230.

Swezey, Sean, Douglas Murray, and Rainer Daxl. 1986. "Nicaragua's Revolution in Pesticide Policy." *Environment* 28(1): 6–36.

Thompson, Karen S., Kathleen M. DeWalt, and Billie R. DeWalt. 1985. *Household Food Use in Three Rural Communities*. Farming Systems Research in Southern Honduras, Report No. 2. Lexington, KY: University of Kentucky Department of Anthropology.

Thrupp, Lori Ann. 1988. "Pesticides and Policies: Approaches to Pest-Control Dilemmas in Nicaragua and Costa Rica." *Latin American Perspectives* 15(4): 37–70.

———. 1991. "Sterilization of Workers from Pesticide Exposure: the Causes and Consequences of DBCP-Induced Damage in Costa Rica and Beyond." *International Journal of Health Services* 21(4): 731–757.

Tierney, John. 1990. "Betting the Planet." *New York Times Magazine*. December 2, 1990, pp. 52–53, 74–81.

Torres Rivas, Edelberto. 1973. *Interpretación del Desarrollo Social Centroamericano*. San José, Costa Rica: EDUCA.

UNDP (United Nations Development Programme). 1990. *Human Development Report: 1990*. Washington, D.C.: UNDP.

———. 1992. *Human Development Report: 1992*. Washington, D.C.: UNDP.

UNFPA (United Nations Population Fund). 1991. *Population, Resources, and the Environment: The Critical Challenges*. New York, NY: UNFPA.

UNICEF (United Nations Children's Fund). 1991. *Country Programme Recommendation for Honduras*. E/ICED/1991/P/L.11. 28 January.

UNWCED (United Nations World Commission on Environment and Development). 1987. *Our Common Future*. Oxford: Oxford University Press.

USDC (United States Department of Commerce). 1956. *Investigations in Central America*. United States Department of Commerce, Bureau of Foreign Commerce. Washington, D.C.: USGPO.

USAID (United States Agency for International Development). 1978. *Agricultural Assessment of Honduras.* Vol. I and II. Tegucigalpa, Honduras: USAID/Honduras.

————. 1982. *Honduras: Country Environmental Profile.* McLean, VA: JRB Associates.

————. 1985. *Environmental Assessment of the Small Scale Shrimp Farming Component of the USAID/Honduras Rural Technologies Project.* Gainesville, FL: Tropical Research and Development, Inc.

————. 1989a. *Considerations for the Agricultural Sector in Honduras.* Tegucigalpa, Honduras: USAID/Honduras.

————. 1988. *Credit Policy in Honduras in the Context of Macroeconomic Constraints.* May. Tegucigalpa, Honduras: USAID/Honduras.

————. 1989b. *Strategic Considerations for the Agricultural Sector in Honduras.* Draft copy. USAID/Honduras: Office of Agriculture and Rural Development.

————. 1989c. *Environmental and Natural Resource Management in Central America: A Strategy for AID Assistance.* Prepared for the Latin American & the Caribbean Bureau by the Regional Office for Central America and Panama.

————. 1989d. *Agricultural Crop Diversification Export Project.* Washington, D.C.: USAID/Experience, Inc.

————. 1990. *Agricultural Sector Strategy Paper.* Tegucigalpa, Honduras: USAID/Honduras.

————. 1991a. *Congressional Presentation, FY 1991,* 166. Washington, D.C.: USAID.

————. 1991b. *Economic Assistance Strategy for Central America, 1991 to 2000.* Washington, D.C.: USAID.

————. 1992. *Congressional Presentation, FY 1992.* Washington, D.C.: USAID.

————. 1993. *Congressional Presentation, FY 1993.* Washington, D.C.: USAID.

USAID/FEPROEXAAH. (United States Agency for International Development/Honduran Federation for the Promotion of Agricultural Exports). 1988. *Study of the Honduran Shrimp Industry.* Tegucigalpa, Honduras: USAID/Honduras.

————. 1989. *Plan de Desarrollo del Camarón en Honduras.* Tegucigalpa, Honduras: USAID/Honduras.

USDA (United States Department of Agriculture). 1959–85. *World Indices of Agricultural and Food Production, 1950–84.* Washington, D.C.: USDA.

————. 1985. *World Indices of Agricultural and Food Production, 1950-84.* Washington, D.C.: USDA.

————. 1987. *Livestock and Meat.* Foreign Agriculture Circulars. Washington, D.C.: USDA. Cited in Leonard 1987: 216–217.

Valverde, Victor. 1986. "Nutritional Status in Central America and Panama." Paper presented at the conference, *Health and Nutrition in the Americas.* University of Kentucky, Lexington, KY, April.

Valverde, V., W. Vargas, I. Rawson, G. Calderón, R. Rosabal, and R. Gutiérrez. 1975. "La Deficiencia Calórica en Preescolares del Área Rural de Costa Rica." *Archivo Latinoamericano. Nutrición* 25(4): 351–361.

Villanueva, B. 1968. *Institutional Innovation and Economic Development: Honduras, a Case Study.* Ph.D. dissertation. University of Wisconsin, Madison, WI.

von Braun, Joachim, David Hotchkiss, and Marten Immink. 1989. *Nontraditional Export Crops in Guatemala: Effects on Production, Income, and Nutrition.* Research Report 73. Washington, D.C.: International Food Policy Research Institute.

Warman, Arturo. 1981. *We Come to Object: the Peasants of Morelos and the National State.* Baltimore, MD: The Johns Hopkins University Press.

Weir, David and Mark Shapiro. 1981. *Circle of Poison: Pesticides and People in a Hungry World.* San Francisco, CA: Institute for Food and Development Policy.

West, Robert C. 1959. "The Mining Economy of Honduras During the Colonial Period." *Actas del XXXIII Congreso Internacional de Americanistas* 2: 767–777.

White, Robert A. 1972. "The Adult Education Program of Acción Cultural Popular Hondureña: An Evaluation of the Rural Development Potential of the Radio School Movement in Hnduras." Department of Anthropology and Sociology, St. Louis, MO: St. Louis.University. Mimeograph.

———. 1977. *Structural Factors in Rural Development: The Church and the Peasant in Honduras.* Ph.D. dissertation. Cornell University, Ithaca, NY.

Whiteford, Scott and Anne E. Ferguson. 1991. "Social Dimensions of Food Security and Hunger: An Overview." In *Harvest of Want: Food Security in Central America and Mexico,* edited by Anne Ferguson and Scott Whiteford. Boulder, CO: Westview Press, 1–21.

WHO (World Health Organization). 1990. *Public Health Impact of Pesticides Used in Agriculture.* Geneva: World Health Organization.

Williams, R. 1986. *Export Agriculture and the Crisis in Central America.* Chapel Hill, NC: University of North Carolina Press.

———. 1991. "Land, Labor, and the Crisis in Central America." In *Harvest of Want: Food Security in Central America and Mexico,* edited by Anne Ferguson and Scott Whiteford. Boulder, CO: Westview Press, 23–44.

Wilson, Charles. 1947. *Empire in Green and Gold: The Story of the American Banana Trade.* New York, NY: Holt.

Wolf, Eric. 1972. "Ownership and Political Ecology." *Anthropological Quarterly* 45(3, July): 201–205.

Woodward, Ralph Lee, Jr. 1976. *Central America: A Nation Divided.* New York, NY: Oxford University Press.

World Bank. Various years. *World Development Report.* New York, NY: Oxford University Press.

Wortman, Miles L. 1982. *Government and Society in Central America 1680–1840.* New York, NY: Columbia University Press.

WRI (World Resources Institute). 1992. *World Resources: 1992–93.* New York and Oxford: Oxford University Press.

Wright, Angus. 1986. "Rethinking the Circle of Poison: The Politics of Pesticide Poisoning Among Mexican Farm Workers." *Latin American Perspectives* 13(4): 26–59.

———. 1990. *The Death of Ramon Gonzalez: The Modern Agricultural Dilemma.* Austin, TX: University of Texas Press.

Zuniga, Melba. 1978. *La Familia Campesina.* Tegucigalpa, Honduras: Instituto de Investigaciones Socioeconomicas. Mimeograph.

About the Book
and Author

This book is about interconnections—those among the historical, geographic, demographic, social, economic, and ecological aspects of development—as well as how Central Americans struggle with the interplay of increasing poverty and environmental degradation. Centering on the case of southern Honduras and expanding to include the Central American region, Susan Stonich's analysis employs an integrative approach that builds on a strong and varied methodological foundation to encompass both political economy and ecology.

Stonich examines the systemic linkages among the dynamics of dominant development models and associated patterns of capitalist accumulation, regional demography, rural impoverishment, and environmental decline. By casting the discussion against the backdrop of southern Honduras, she presents a powerful historical record of how larger sociopolitical communities impact individuals and the natural environment and how, in turn, people respond. She charts the destiny of peasant groups within the dynamics of contemporary capitalism, recognizing that the fates of the peasantry and the natural environment are intimately linked. Stonich's study contributes to an improved understanding of the complex interrelationships between social processes and environmental degradation, offering a timely and pertinent comment on one of the most serious modern challenges.

Susan C. Stonich is associate professor in the Department of Anthropology and the Environmental Studies Program at the University of California at Santa Barbara.